Imagining
the Internet

Imagining the Internet

Personalities, Predictions, Perspectives

JANNA QUITNEY ANDERSON

ROWMAN & LITTLEFIELD PUBLISHERS, INC.
Lanham • Boulder • New York • Toronto • Oxford

ROWMAN & LITTLEFIELD PUBLISHERS, INC.

Published in the United States of America
by Rowman & Littlefield Publishers, Inc.
A wholly owned subsidary of The Rowman & Littlefield Publishing Group, Inc.
4501 Forbes Boulevard, Suite 200, Lanham, Maryland 20706
www.rowmanlittlefield.com

P.O. Box 317, Oxford OX2 9RU, UK

British Library Cataloguing in Publication Information Available

Library of Congress Cataloging-in-Publication Data
Anderson, Janna Quitney, 1955–
  Imagining the internet : personalities, predictions, perspectives / Janna Quitney Anderson.
    p. cm.
  Includes bibliographical references and index.
  ISBN 0-7425-3936-9 (cloth : alk. paper) — ISBN 0-7425-3937-7 (pbk. : alk. paper)
  1. Internet—Social aspects—Forecasting. 2. Internet—Public opinion—History. 3. Information technology—Social aspects—Forecasting. 4. Computer networks—Social aspects—Forecasting. 5. Forecasting—History—20th century. I. Title.
  HM851A63 2205
  303.48'33'0112—dc22

2005008431

Printed in the United States of America

♾™ The paper used in this publication meets the minimum requirements of American National Standard for Information Sciences—Permanence of Paper for Printed Library Materials, ANSI/NISO Z39.48-1992.

Contents

Foreword

Prescient predictors' statements enrich us all on many levels; share your vision, too

Around the turn of the century, when I began my work as director of the Pew Internet & American Life Project, it was my task to find new approaches for the study of this new communications medium. When I decided to sponsor a research project to collect Internet predictions, my motives were pure, but my expectations were sensationalist. I knew it would be valuable to examine what the creators of the Internet and social commentators had said in the early 1990s about the probable impact of the Internet. Prominent thinkers often shape the direction of technological development. Yet, I also expected that any serious canvassing of early predictions would yield a bumper crop of howlers.

I thought that somebody—probably lots of somebodies—had forecast the future of the Internet in ways as spectacularly silly as bygone-era mispredictors. Famously, a Western Union official crafted an internal memo in 1876 after the telephone was demonstrated, sniffing that, "This telephone has too many shortcomings to be seriously considered as a means of communication." And Henry Morton, the president of the Stevens Institute of Technology, watched the first demonstration of Thomas Edison's electric light, in 1879, and declared: "Everyone acquainted with the subject will recognize it as a conspicuous failure."

My hope for "gotcha" examples of flat-wrong Internet predictions was dashed the first time I looked at the fruits of the predictions research done by Elon University's Janna Quitney Anderson and her colleagues. It was striking to see how prescient so many technologists and other analysts of the early 1990s were about the coming impact of the Internet. They saw the future and their vision clearly shaped it.

When one reads the material from this early-Internet era one must applaud the foresight of many people, including Tufts University's Professor Nan Levinson, who said in a 1992 paper posted on the Electronic Frontier Foundation site: "Widespread and fairly allocated computerized resources can offer: increased citizen participation in and oversight of government affairs; assembly, organizing, and debate unrestricted by geographical distances or boundaries; decentralized decision-making; a challenge to news and publishing monopolies; rapid international exchange of information; and individually-tailored, focused information to combat the information glut that interferes with communication."

This and more than 4,000 additional predictive statements made by 1,000 Internet stakeholders and skeptics are now gathered in Elon University's Pew Internet–funded study of public speeches, articles, and books that appeared between 1990 and 1995. Dozens of these predictions are put into context in this book, which looks at the future through an analysis of the past.

It is somewhat difficult after becoming immersed in these insights to remember that Internet communication began with the utmost diffidence. Indeed the first events involved a computer crash and unmemorable twaddle.

The first attempted transfer of information packets in 1969 was not launched with the same portentous thunder as was Samuel Morse's telegraph message in 1844: "What hath God wrought?" Rather, Charley Kline, an engineer at the University of California at Los Angeles, froze his computer in 1969 when he began typing "L-O-G" (on his way to "L-O-G-I-N") to start the file-transfer program. Programmers fixed the glitch quickly, and the file sharing began.

E-mail arrived in 1971, and again there was no self-puffery in the inaugural text. Indeed, there was not even an effort to match the practical tone of Alexander Graham Bell's initial phone call in 1876, "Mr. Watson, come here; I want you." An engineer named Ray Tomlinson sent a test message from one computer to another that was sitting less than five feet away. He cannot remember exactly what he typed, but he thinks it was probably "QWER-TYUIOP," and it showed up on the neighboring screen. It was a milestone moment that was not self-evident. Tomlinson did not think much of his lovely little hack that allowed people using computers at remote sites to write electronic notes to each other. Today, the one thing for which Tomlinson is best remembered is that he is the originator of the use of "@" as the locator symbol in electronic addresses.

When I first talked to Elon University scholars about my idea for a predictions research project, I think I made my mistaken guess about the number of wrong-headed predictions we would find because I hatched this idea in mid-2000, at the height of dot-com mania. The air then was filled with moronic overblown prophecies about the course of the Internet. My hunch now is that the craziest things predicted about the Internet were made by the *second wave* of commentators; those who really believed that the concept "the Internet changes everything" meant that they could exploit the revolution for commercial purposes. The people who were widely quoted and aggressively published in the period between 1996 and 2000 were the ones who said transcendently dumb things about what people would do online and fabulously overpredicted the consequences of Internet adoption by the masses.

Would that they had heeded the 1994 words of technology writer Vic Sussman in his *U.S. News & World Report* article "The Internet Will Gain Popularity, Problems":

> Many of the commercial interests scrambling to mine gold in cyberspace will crash and burn, victims of their own inexperience in this unique world. . . . A burgeoning Internet will also force society to confront traditional notions of free speech and intellectual property in new, sometimes uncomfortable ways. What happens to the idea of "community standards" when almost anyone with a computer and modem can become a self-publisher with global distribution? A flood of new users may spark an epidemic of busy signals, worries about data security and a potential boom in computer crime. As the Net throws open its cloistered universe, the real world—goblins and all—will come rushing in.

The gains to be made by studying people's predictions about the future of networked communications are so compelling that the Pew Internet Project worked with Elon to develop the Predictions Database further. In addition to the "Early '90s" section, which includes more than 4,000 statements made by 1,000 Internet stakeholders and skeptics, we have invited Internet experts and the general public to add new contributions. The "Experts Survey" section of the Predictions Database site includes forecasts made in 2004 by technology stakeholders Elon University and Pew Internet surveyed in a look toward 2014. The "Share Your Vision" section allows anyone (we encourage you and your friends to join in) to venture his or her own forecast. All of these resources are online at www.elon.edu/predictions.

These efforts to assess the past and look into the future are part of the ongoing research of the Pew Internet & American Life Project. We continue to conduct surveys about the future of communication and information technologies as part of our goal to track the cultural changes wrought by the Internet. In particular, we hope that this material will be useful to: scholars who wish to assess the distance we have come; journalists who are trying to figure out where we are now; government, industry, and nonprofit officials who want to build the Internet of the future; and people of all walks of life who must learn to recognize the coming complexities of their networked world.

The research here grows out of the mission of the Pew Internet Project, which was born of the idea that the impact of the Internet is testable. Rebecca Rimel, president of the Pew Charitable Trusts, was struck by the fact that many of the debates about the impact of the Internet in the late 1990s were taking place without reference to data and basic social-science research. She and the board of trustees of the foundation had the foresight to believe that a research project could play a useful role in the disputes by providing nonpartisan facts and analysis about the social impact of people's Internet use.

The foundation provides generous support for a research agenda that monitors Americans' use of the Internet and focuses on several aspects of Internet use that are not major concerns of the many proprietary research firms that concentrate on e-commerce. Those areas of emphasis include how people's Internet use affects their interactions with family, friends, and others; their involvement with various communities; their health care; their educational experiences; their civic and political lives; and their workplaces.

By 2004, the Pew Internet & American Life Project had produced nearly 100 reports about the social impact of the Internet. Of all the things we have done, nothing has taught us as much and made us think as expansively as this effort. We are very grateful to Janna Quitney Anderson, her Elon University colleagues, and their students for their dogged, shoe-leather work in tracking these predictions down—and for the synthesis and analysis Anderson shares in this book.

On our best days at the Project, we hope we produce material that is valued as a "public good" by other researchers and those who want to understand the role of the Internet in American life. The creation of the Predictions Database, the thoughtful insights offered by experts willing to share new predictions, the contributions of the public at large as people submit their visions, and the fund of knowledge available in this book are the kind of public good we were funded to create.

Lee Rainie
Washington, D.C.

Introduction

Why it is important for you to understand networks and their influence in history and your life

The world is made of nested networks, from those at the invisible, subatomic level to those plainly visible in nature and in human-created tools and technologies.

Change is accelerating as networked communications revolutionize our social, political, and economic networks. In 1969, the first computer network was made up of four sites, and by 1977 it included 50. By 1994, the network of computer networks had 4 million host computers. This was the year in which microprocessors came to outnumber humans. By the first decade of the 21st century, computers had become ubiquitous and connectivity had become universal in all economically advanced communities. People could communicate remotely, asynchronously, and indirectly.

Pervasive information networks are being interwoven invisibly into our environment. Many of these are high-performance sensors, known as MEMS (for microelectromechanical systems). By 2004 there were already at least five of these devices per person in the United States, and the market for MEMS had grown to become a $10 billion industry. Mark Weiser foretold this ubiquity of stealthy smart devices in his 1991 *Scientific American* article, "Future Computers." He wrote: "The most profound technologies are those that disappear. They weave themselves into the fabric of everyday life until they are indistinguishable from it." By 2005, millions of embedded micro devices already served in thermostats, pollution detectors, consumer products, highways, trucks, and in devices we can have implanted in our pets to prevent their loss or in ourselves to provide details about life-threatening health problems to those who wish to help us.

Two of the current concerns expressed by experts as we all become linked in myriad ways are the potential for the end of privacy and the end of property. As information technology is quickly becoming an unnoticeable part of everyday objects, the era of privacy has come to an end. Surveillance cameras and sensors are everywhere. Radio frequency identification devices or RFIDs—miniscule signal-transmitting tags that can be easily embedded—are being used to track much more than the migratory habits of birds. It's not just the delivery companies such as United Parcel Service using them. They are already attached to more items than you may realize; think twice before taking that souvenir hotel towel home from your next vacation. Thanks to the rise of digital networks, music and the written word are no longer intellectual properties that can be easily protected by copyright ownership. As some individuals and corporations battle to retain a degree of control over their creations, others are glorying in the opportunity to share their work for free.

A prime concern of our networked present and future is the maintenance of a steady source of energy and a model of consumption that is ecologically sustainable. The human network has become dependent on transportation networks and communications networks of growing sophistication. All of these human-created systems come to a standstill when the power grid—another network—shuts down, as people on the U.S. East Coast found in August 2003. One hope is that intelligent networks can help us find ways to produce power in a clean, low-impact format, and that they may also lead us to establish a reasonable sustainable development strategy that respects the planet and all of its creatures.

One final, overarching concern is network control. Information is power, and communications networks at this point are physical entities that can be dominated. It is vital that networks be operated in such a way that free speech is the rule; there can be no network monopoly owned by commercial, political, or ideological interests.

Far into the future will come advances that could lead to machine superiority—not necessarily a bad thing if it's handled correctly, since many experts predict this could bring the end of work, allowing people to pursue their dreams. Today's nanotechnologists are figuring out ways in which intelligence could be placed in molecular-size devices. This is worrisome and requires attention: Some theorists have even gone so far as to postulate that artificial-intelligence development may eventually evolve into a "godmind" or an invisible intelligence that eliminates the need for humankind.

This book is a potpourri of peeks into the future and the past, linking personalities and technologies in an informative package showcasing the Internet while putting communications history and its influences on social, political, and economic systems into some perspective.

The catalyst for the research study that led to this book is Lee Rainie, founding director of the Pew Internet & American Life Project. He thought it would be interesting to look through published records of the early 1990s to see what people were predicting the Internet would become. His idea, and the financial backing of Pew Charitable Trusts, led a group at Elon University to the development of a database of a thousand voices making thousands of Internet predictions. In our research, we sought out and mined predictive data, reading millions of words in thousands of articles, transcripts and chapters, and finding forecasts in nearly 500 publications of the writings, presentations, speeches, and interviews of people who had something important to say about the Internet between 1990 and 1995. This book includes and was inspired by the voices of the stakeholders and skeptics now recorded in the Elon University/Pew Internet & American Life Predictions Database, which can be found at www.elon.edu/predictions.

Our social, political, and economic systems are generally still wrapped up in the rules and institutions of the industrial age. This must change. As we go forward, we must think several jumps ahead. Decisions being made today are making your future. Inform yourself and become involved; let's all think together and work together to make it the best possible future.

The Internet at the Forefront

1990 through 1995 were revolutionary, with changes surpassing any previous stretch in communications history

How does the arrival of the Internet compare to the introduction of other new communications tools? It took 38 years for radio to get a market of at least 50 million users; it took television 13 years to get to the 50-million-users mark; and it is estimated that it took just four years for the public Internet to achieve 50 million users.

The growth of digital communications dwarfs that of the printed page, radio, and television. In addition to providing an overview of the social history of communications technology, this book describes in great detail what people were saying about the Internet during its defining years of public acceptance: 1990 to 1995. This is considered the "awe" stage of the medium, when technological breakthroughs made it useable and affordable and it caught the fancy of the general public. Peter Deutsch, the developer of Archie, an Internet index of the early 1990s, said 1990 brought the turning point in the way people perceived and made use of the Internet. "The Internet is no longer just an interesting experimental testbed," he wrote in a 1992 issue of *Computing Systems* magazine. "The Net is now a daily tool for hundreds of thousands of people who couldn't tell an IP packet from a burst of tty line noise."

The arrival of digital networked communications has been referred to as the "Digital Big Bang" of the 20th century. Humankind went from the use of one computer in 1946 to a few thousand in the 1960s to untold millions at the start of the new millennium, and (with the incursion of digital sensors and other networked devices that aren't used directly by humans) an expectation of a few hundred billion in the next 50 years. How far did networked communications move in just 20 years of that span? Between 1980 and 2000, we went from being able to transmit less than one page of information per second on a copper phone wire to the ability to swiftly transmit more than 90,000 volumes of an encyclopedia on a hair-width strand of optical fiber.

WHY WE SEE VALUE IN REMARKS FROM THE PAST ABOUT THE FUTURE

Over the past several centuries, the arrival of such world-altering communications technologies as the printing press (1453), radio (1896), and television (1927) caused social commentators, business people, and politicians of those times to predict what might come to pass due to changes wrought by such new devices. Their aim in making predictive statements was to prepare their world—to brace it for inevitable economic, political, and social adjustments. As Ithiel de Sola Pool, an esteemed researcher of the 20th century, so aptly put it in his 1983 study *Technologies of Freedom*, "These technologies caused revised conceptions of man's place in the universe."

Pool's pioneering work in the study of social science came in the decades after he helped Harold Lasswell and Nathan Leites complete their research on Nazi and Communist propaganda during World War II. He spent 30 years at MIT investigating the effects of communications technology. His hundreds of influential research works touched nearly every field of communications, society, and politics, including: *Newsmen's Fantasies, Audiences, and Newswriting* (1959); a chapter on public opinion in *The Handbook of Communication* (1973); *Forecasting the Telephone: A Retrospective Technology Assessment* (1983); and his best-known work, *Technologies of Freedom: On Free Speech in an Electronic Age* (1983), an incisive study of the ways in which communications technologies transform social, political, and economic life.

Pool's groundbreaking research into the "small-world phenomenon" outlines the close proximity each human has to others because of the crossovers of social networks. Pool was a pioneer in identifying and describing these "contact networks"; he expected developments in communications technologies to make these even more widespread. At the time he did his research into how people relate in cascading chains of influence, he had no idea that a playwright would later turn his concept into a Broadway success called *Six Degrees of Separation,* nor that popular culture of the 1990s would adapt the idea yet again into a witty entertainment called "Six Degrees of Kevin Bacon."

His study of human networks and communications networks gave Pool a clear perspective few others of his time could claim. In a 1983 book chapter titled "Development of Communication in the Future Perspective," published in *The Human Use of Human Ideas,* Pool said: "People who think about social change in traditional terms cannot begin to imagine the changes that lie ahead. Conventional reformers cast their programs in terms of national policies, or in terms of laws and central planning. But in the end, what will shape the future is a creative potential that inheres in the new technologies."

Although he died in the mid-1980s, Pool's work was influential in the decade that followed, when vital decisions were being made regarding freedom on the Internet. Policymakers and researchers were inspired by his *Technologies of Freedom.* Pool projected that interconnected computers would build a free-wheeling, wide-open communications structure ("the largest machine that man has ever constructed—the global telecommunications network; the full map of it no one knows; it changes every day") that would be questioned by regulators fearing the challenge to the economic and political status quo. He warned that a positive future would be delayed if regulators chose to interfere. Activist and Internet pioneer Stewart Brand wrote in a 1989 Global Business Network online review, "I've seen this book convert liberals away from government control of broadcast media toward a guided marketplace approach. . . . I've seen technology skeptics . . . begin to get a gleam in their eye."

Certainly the Internet and the dawn of networked digital communications have altered humankind's conception of its place in the universe more significantly than any previous tool. It is a public utility that allows anyone, anywhere, to access and share any information, visual and aural, nearly instantly, and at relatively little cost. It has already caused change, and it could be the precursor to a networked artificial intelligence with unlimited potential for good and evil.

The prophets who seek to foresee the consequences of a new technology often do so in the hope of making a profit. Many others are motivated by the ideal that better social choices can be made if the coming impact of a new tool can be accurately pre-assessed. An observance of what stakeholders and skeptics were saying at the dawn of a new communications age is as revealing as a study of past wars and the making of the peace that followed. This book is a look at the potential future of networks and an examination of the social, political, and economic history of the Internet, as seen through the eyes of the stakeholders and skeptics of its early boom years. Let's begin our examination of the revolution with a scene-setting history of Internet in the early 1990s.

BERNERS-LEE ENQUIRES AND THE WEB IS BORN

Several key elements—including the World Wide Web, Mosaic, and Java—came together to help networked digital communications explode on the 1990s scene as it never had before.

When physics researcher Tim Berners-Lee had the initial concept for a software program that would make hypertext links possible, he named it Enquire, in honor of "Enquire Within Upon Everything," a book of Victorian advice he'd grown up with in his parents' house in London. His parents were mathematicians who had helped program the world's first commercial stored-program computer, the Manchester University Mark I.

Berners-Lee graduated from Oxford in 1976 and wrote Enquire in the early 1980s to aid his efforts as a software consultant at CERN (Conseil Europeen pour la Recherche Nucleaire), near Geneva, Switzerland. His work required information sharing. The basic Enquire code he wrote in Pascal computer language allowed him to link work files internally and externally. When he left CERN for a time, he left the Enquire source code behind and it was lost. He rejoined CERN in 1984 and began to recreate Enquire hoping to combine external linking with hypertext and interconnectedness: He dreamed of linking all of the world's computers so they could share data.

Berners-Lee's part-time networked computing research efforts over the span of a decade led to the development of the first HTML (hypertext mark-up language) source code in 1990. He pitched the idea at the European Conference on Hypertext Technology in September of that year, and then wrote the HTML code and server code for what he called the "World Wide Web." By Christmas Day of 1990, he'd programmed browsers on two computers—his and CERN colleague Robert Cailliau's—to communicate over the Internet through the use of the info.cern.ch server. He'd found that it was necessary to be able to demonstrate his idea in order to sell it—just as Samuel Morse discovered that Congress would not fund his telegraph until he demonstrated a working model in 1842. Berners-Lee and Cailliau used their setup to follow links to hypertext-transfer-protocol servers and file-transfer-protocol-based Internet news and newsgroups and thus demonstrated the value of the Web.

Berners-Lee's World Wide Web didn't actually appear online and come into use until 1991. "The idea of universality was key," he explained in his 1999 book *Weaving the Web.* "The system should not constrain the user; a person should be able to link with equal ease to any document wherever it happens to be stored."

INTERNET COMMITTEES PROLIFERATE

1990 was also the year that the Advanced Research Projects Agency Network (ARPANET), initially a military-financed project and at the time mostly used by researchers, was decommissioned after 20 years of operation; the National Science Foundation (NSFNET) backbone—at least 25 times faster than ARPANET—took over and further democratized the network. Researchers and government planners began to see the need to form committees to seriously assess the ramifications of the rollout of this new technology. The National Research and Education Network (NREN) emerged during the George H.W. Bush administration in 1990 as part of the High-Performance Computing Program. It was expanded during the Clinton administration to include funding for a National Information Infrastructure—meant to advance networked computing options in schools, libraries, and hospitals.

From its earliest days, collaboration and close critique were keys to the development of networked computers. ARPANET and NSFNET were built by computer pioneers in a market-free environment, supported by funding from the public. With no commercial interests involved at its start, the Internet was developed in a cooperative, experimental environment. In 1990, government and communications industry leaders met at Harvard University, where they hammered out a plan described in the document "Commercialization of the Internet: Summary Report." By 1993, the U.S. government agreed to turn continuing network development over to the private sector, announcing a privatization plan called the "National Information Infrastructure Agenda for Action."

It is difficult to explain how the management of the Internet's establishment and architecture came about without introducing an alphabet soup of organizing bodies. ARPA program manager Vinton Cerf had convened the Internet Configuration Control Board (ICCB) in 1979 to guide the technical evolution of the Internet. It was reorganized and renamed the Internet Activities Board (IAB) in 1983. Over time, representatives from government organizations including ARPA, the National Science Foundation, NASA, and the Department of Energy formed the Federal Research Internet Coordinating Committee (FRICC), which supported the IAB. In the ensuing years, an impressive array of additional Internet-oriented organizations popped up on the scene. In 1990, FRICC was reorganized and created the Federal Networking Council (FNC) to be the official governing body for coordinating all of the agencies supporting the Internet. In addition, an international Coordinating Committee for Intercontinental Research Networks (CCIRN) was created to establish cooperative planning between European and North American networking groups. The IAB continued as a worldwide organization at this point, setting Internet standards, working as an international technical policy liaison, and accomplishing strategic planning.

The Internet Society, known as ISOC, was established in 1991 by the Center for National Research Initiatives and the IAB. Cerf was the first president when the society, which was established as an international body to oversee the IAB and its Internet Engineering Task Force and Internet Research Task Force, got its start in 1992.

THANKS TO MOSAIC, A NATION GOES NUTS FOR THE NET

In 1991, thanks to the ease-of-use on the horizon spurred by Berners-Lee's Web, Internet Service Providers (ISPs)—businesses that allowed people to use their telephone lines to get access to use of the Internet—began gaining popularity. Such services included CompuServe and Prodigy. The first user-friendly network interface was named "Gopher," after the sports mascot at the University of Minnesota, where it was created in 1991. Gopher was extremely limited in comparison with tools soon to come, but it was the best thing yet to emerge in Internet communication, and it was nearly universally adopted.

The public introduction of the World Wide Web in 1993 laid the groundwork for boom times for the Internet. The annual rate of growth for the digital communications network that year was estimated at 341,634 percent by the Internet Society. New networks were connecting to the system at a rate of one every 10 minutes—an astonishing figure at the time. By midyear the Internet included more than 15,000 networks and 2 million computers.

Thousands of newspaper and magazine articles about the Internet popped up in the popular press in the first months of 1993. Despite the fact that the digital network had been used by the research community for decades, it seemed to the general public as if it had appeared out of nowhere. The World

Wide Web made information sharing easier, and Mosaic, which debuted in April of 1993, helped people find a better way to get to that information.

Marc Andreessen had developed the idea behind Mosaic in 1991–1992 as an undergraduate student at the University of Illinois, where he worked with the staff of the National Center for Supercomputing Applications. He enlisted the help of a group of students and NCSA staffers to create this much-improved Internet browser. Mosaic later became Netscape, and over the following decade it inspired Microsoft's Internet Explorer, Apple's Safari, and many ensuing user-friendly Internet-navigation systems.

Because of the excitement generated by the revolutionary Mosaic software program, hundreds of thousands of people who had never used the Internet began to see the benefits of sharing content—in e-mails or on Web pages—on the worldwide network. The system began to slow down as online traffic increased. The engineers and scientists who were working to keep the network up and running had to scramble to figure out how they could meet the daunting challenge of accommodating an exponential growth rate.

Before Mosaic hit the screen in 1993, Gopher was the tool of choice for most Internet communicators. Anthony Rutkowski of the Internet Society estimated the annual rate of growth of Gopher traffic at 997 percent in the fall of 1993. Nowadays, if you know Gopher as something other than a rodent or that goofy guy on the vintage TV series *The Love Boat,* you are probably an Internet old-timer.

Yes, 1993 was the year it all broke loose. *Wired* magazine, the witty, wise, mainstream technology publication that helped illuminate issues and prod politicians, got off the ground in January. On March 2, Bill Clinton wrote the first Internet mail message ever sent by a U.S. president. The White House put up its own Web site, and announced that any citizen could send e-mail to the president or vice president. By May, National Public Radio had begun to broadcast some programming on the Internet. One of the nation's richest and most prolific writers of fiction, Stephen King, published a short story online that fall—*before* he published it in print.

In the first issue of *Wired,* founding editor and publisher Louis Rossetto wrote: "The digital revolution is whipping through our lives like a Bengali typhoon . . . the computer 'press' is too busy churning out the latest PCInfo-ComputingCorporateWorld iteration of its ad sales formula cum parts catalog to discuss the meaning or context of social changes so profound their only parallel is probably the discovery of fire." Vital issues that had been discussed in small groups online were made more public by the magazine, and its content reflected the concerns of those knowledgeable about the new technology and its positive and negative potential, covering economic systems, copyrights, the media, schools, libraries, warfare, and privacy, among other things.

COMMERCIALISM BEGAN TO SOAR IN 1994

In 1994, the year the ARPANET—the original Internet—celebrated its 25th anniversary, the first online bank (First Virtual) opened and so did the first online shopping "malls." Radio stations began broadcasting on the Internet. *Wired* magazine spun off an online product that was aimed at expanding into new territory from the print version—called *HotWired*, it also ran some of the first online banner advertising (for AT&T and Zima). This was the year that the Arizona law firm Canter & Siegel was the first to generally "spam" everyone on the Internet, sending out a mass e-mail advertising its services—netizens "flamed" (replied with nasty retorts) in return.

The ads on HotWired combined with the spamming incident raised a red flag—purists had warned that if the Internet went public and became easily accessible, runaway commercialism could ruin this unique communications medium. Commercialism had arrived, but it co-existed with the same free-speech content of the early days.

1994 was the year in which William Mitchell of the Massachusetts Institute of Technology wrote the fascinating predictive book *City of Bits*. It did not appeal to the mass audience at the time, but Mitchell convincingly outlined a future in which buildings are smart and even your underwear is networked:

> We will all become mighty morphing cyborgs capable of reconfiguring ourselves by the minute—of renting extended nervous tissue and organ capacity and of redeploying our extensions in space as our needs change and as our resources allow. Think of yourself on some evening in the not-so-distant future, when wearable, fitted, and implanted electronic organs connected by bodynets are as commonplace as cotton; your intimate infrastructure connects you seamlessly to a planetful of bits, and you have software in your underwear. It's eleven o'clock, Smarty Pants; do you know where your network extensions are tonight?

Scientists and engineers who were keeping the Internet running were concerned about all of the new traffic online at this point in time. It was slowing the rate at which information could be exchanged. How would they keep things going? If everyone was to jump onboard, the entire system might crash.

They worked together to find a way to squeeze more out of a network being inundated with more users doing more things. Internet Society leader Cerf predicted in September 1994 that by 2000 there would be 300 million Internet users; by November 1994 Cerf had upped his estimate to 400 million users.

One futurist saw the sunny side to come, correctly foretelling the eventual growth of the bandwidth necessary for the Internet to continue without a glitch: "The tide is now gathering toward a crest," wrote technology consultant George Gilder in the Forbes ASAP article "The Bandwidth Tidal Wave." "It will be possible to carry 2.4 gigahertz (billions of cycles per second) on each wavelength stream. That would add up to more than 1,700 gigahertz on every fiber thread."

SUN MICROSYSTEMS' JAVA ADDS JIVE IN '95

In 1995, the National Science Foundation's NSFNET withdrew from network control and the main backbone traffic in the United States began to be routed through interconnected providers, opening the door for commercial development. The Vatican, the U.S. House and Senate, and the Canadian government built and launched online sites. In late May, Sun Microsystems introduced Java, the software that enabled Web pages to move, groove, and come to life. The acquisition of domain names had become so popular that what was formerly free now cost $50 per year.

Over the next few months, RealAudio introduced streaming-audio ability to the net, and Mosaic wunderkind Marc Andreessen's new company Netscape had an incredibly successful initial public stock offering on the NAS-DAQ exchange—shares priced at $28 opened at $70; a lot of people were betting the Internet would have a profitable future.

Two of America's best-known high-tech stars came out with books about the future in 1995. Nicholas Negroponte, a co-founder of MIT's groundbreaking Media Lab and a columnist for *Wired* magazine, published *Being Digital*. Microsoft's founder and chief executive officer Bill Gates published *The Road Ahead*.

Negroponte and Gates both used their books to promote among the general public a positive attitude toward technology. Both projected a future that would bring massive change thanks to the new ways data could be exchanged. Negroponte's book was a witty, approachable summation of thoughts inspired by his knowledge of the Media Lab's technology experiments. Gates's book was a folksy roundup of generally accepted theories regarding the future of digital communications that was packaged in simple terms to reach a wide audience. Both men drew their expertise at least partially from knowledge gleaned from their circles of friends and co-workers, all of whom were at the cutting edge of the field.

Negroponte enthused:

> Bits are not edible; in that sense they cannot stop hunger. Computers are not moral; they cannot resolve complex issues like the rights to life and to death. But being digital, nevertheless, does give much cause for optimism. Like a force of nature, the digital age cannot be denied or stopped. It has four very powerful qualities that will result in its ultimate triumph: decentralizing, globalizing, harmonizing, and empowering. . . . The information superhighway may be mostly hype today, but is an understatement about tomorrow. It will exist beyond people's wildest predictions.

"The Internet is a tidal wave," said Gates. "It will wash over the computer industry and many others, drowning those who don't learn to swim in its waves."

LOOKING TO THE NETWORKED FUTURE

Social commentator William Van Dusen Wishard, author of *Between Two Ages*, remarked in the mid 1990s:

> What kind of future is America going to have based on the promotion of illusion rather than the search for truth? We've substituted image for substance, illusion for truth and then we wonder why our children are having difficulty finding their way in life. What this focuses for us is that technology is a means to an end, not an end in itself. . . . Man does not live by data alone. We live in two worlds—the world of data as well as the world of meaning. The more information we amass, the more power our computers gain, the more essential meaning becomes. Finally we must express in fresh terms the enduring meaning of the American experiment with liberty and responsibility. This is not the responsibility of someone else. It's the task of anyone who believes that America has something great and unique to contribute to the future of the human endeavor.

It is revealing to look back at what forward-looking people were thinking during the "awe" stage of the Internet from 1990 to 1995.

People who make predictive statements of any sort do so at the risk of being mistaken. Of course, mistaken forecasts may hinder society's efforts toward understanding the best uses and potential impact of a new technology. As our world becomes more complex with each passing year, it becomes simultaneously simpler and more difficult to come up with prescient forecasts about the future.

Most of the stakeholders and skeptics of the early 1990s who made predictions about the Internet and its potential impact did so carefully, with knowledge of the successes and failures of those who made such statements regarding earlier technologies, from the pencil to the printing press to the personal computer. This may be the reason that most of their remarks seem to be right on target. What seems obvious to us now was not so clear then; it's a tribute to them that they were so accurate in projecting much of what we have seen so far.

There is one concept on which nearly every 1990s Internet seer was in agreement: The people making decisions about the future of the Internet were making choices that could change the world.

2

From Bonfires and Bongos to the Web

People crave and benefit from connections, spurring communications networks to evolve

While there's no doubt it is revolutionary, the Internet is also simply the next step in humans' ascent of the ladder of communications technology. Pioneers have been out there since the dawn of time, striving to find new and better ways to share information.

Early communities formulated systems of signaling over a distance by fire or by drums, but most communications prior to the 1500s were related by oration or delivered by hand. In the Western world, formally recorded copies of written information were horded by the world's powerful elite up until the 15th century, and much of it was controlled primarily by the Catholic Church, which had by then crowded out many of the ideas and ideals of the classical Greeks and Romans. In the Middle Ages, priests and a spare few scholars were allowed to read, and monks were the people assigned the privilege and task of writing.

German silversmith Johannes Gutenberg changed all this in 1453, introducing the printing press. The development was first welcomed as a boon to the church and society by Catholic leaders, but by 1517 the power of paper and ink in a press broke up the world's first multinational media monopoly when Martin Luther published his "95 Theses," which led by 1534 to King Henry VIII's declaration that Catholicism was illegal.

For every inventive Samuel Morse (telegraph) or Guglielmo Marconi (radio) since then, there have been many who failed or were beaten. And while the struggle over rule-making for and control of the Internet has seemed novel, those familiar with communications history know it is merely an echo of the struggles over earlier technologies such as the telegraph, radio, and television.

FROM INNOVATION TO COMMERCIALIZATION TO REGULATION

All modern networked communications technologies have evolved from the work of many individuals riding a wave of creativity and competitiveness. French historian Fernand Braudel proposed the idea of projective history. He said those who wish to foresee the future can only learn so much from the changes the world has seen in leadership and economies and through wars and peace. Rather, the future can be found by studying the things that do not change; in finding eternal truths we can extrapolate that which is to follow. He said we should scrutinize the fixtures to envision the coming wave.

The developmental years of the telegraph, radio, the telephone, television, and the Internet followed identical plotlines: first came a period of innovation, as inventors struggled to find a way to make their ideas work and then found the backing to finance their practical development and push for public acceptance; next came commercialization, as opportunists and entrepreneurs sought and found—often through a process of trial and error—the angles that would bring financial gain; and, finally, came the grudging acceptance of regulations necessitated by fights over patents, standards, and the avoidance of monopolies.

The innovators tend to be young and often are only able to further their ideas with the help of entrepreneurs and/or political backing. Marconi was 20 when he first pushed his idea of radio; Philo Farnsworth was 20 when he first demonstrated his television; Marc Andreessen had just turned 21 when he developed the first Internet browser, a breakthrough nearly as important as the development of the World Wide Web, which began as an idea in the mind of Tim Berners-Lee when he was 35. Innovators generally build upon the work of other inventors and earlier theorists, finding a better way.

The entrepreneurs and/or political backers who team up with the innovators during the commercialization phase and build on their ideas are generally older and well connected. They step in at an early stage and try to corner as much control of the market and/or financial gain as possible, before competitors join the fray. Their rush to capitalize on the financial and social possibilities offered by a new technology is often joined by pirates—in the mid-1800s, for example, many people stole patented telegraph plans to start their own lines, and pirates of the late 1900s hacked their way into private Internet accounts.

The establishment of common structures, rules, and governing bodies evolves out of concerns and conflicts over property rights and patents, often driving entrepreneurs to grudgingly accept some form of regulation. There is generally a turf war, with each technology developer looking for the most advantageous (commercially profitable) position. They regularly go so far as to request a government investigation of a rival. For instance, Western Union was challenged as a telegraph monopoly in the 1870s; RCA's early radio monopoly was broken up in 1926; and Oracle and Netscape, the pioneering Internet firms, successfully lobbied the government to investigate powerful rival Microsoft, which was brought to trial by the federal government in an antitrust suit in the 1990s. Multiple devices and/or systems associated with each new communications tool are usually developed at the beginning of commercialization, competing for use. For example, the television of the 1920s and early 1930s was delivered through two different systems: one was mechanical and the other electronic. The multiple forms of an innovative technology coexist for a time until consumers adopt the one option offering the best quality, ease of use, or most economical cost. Standardization results when consumers flock toward the most attractive of the alternatives available. In the United States, firms often start up industry associations or arrange summits at which representatives of the various competing companies work toward standards or operating agreements. Government standards or regulations often result because those in an industry can't come to agreement. An example is the establishment of the 525-line/30-frame, black-and-white television system in the United States; it took government and industry groups six years to arrive at that standard, which was announced in 1941. The United States and other nations have often varied in this regulatory stage of the development of communications technologies, with Americans following a free-market approach and Europeans tending toward more government ownership and/or stricter government control. When Marconi's wireless-wonder radio came on the scene, officials in Russia, Germany, and France imposed constraints and standards on the new medium—much tougher restrictions than those in the

United States. The same pattern was followed in the early years of television. The Internet can be more difficult to monitor and regulate, although some nations' governments today are doing so with varied levels of success.

Taking a closer look at the history of the telegraph, radio, the telephone, and television will bring the similarities in their invention, dissemination, and regulation to that of the Internet into better focus.

THE TELEGRAPH

The Innovation: Morse Code Followed the Copper-Pots-and-Clocks System

Englishman Stephen Gray discovered in 1729 that electricity could be conveyed across wires. This inspired a number of inventors to seek ways to use electricity. The concept behind the telegraph was developed by two French brothers, Claude and Rene Chappe, in 1791. They sent messages by sound by using a system of copper pots and coded them through the use of synchronized clocks. It was rough system, but by 1798 such telegraph lines were being used to transmit news of French military victories.

Other European nations began experimenting by building wooden towers and using different codes and symbols, and inventors in Europe and the United States sought ways to send coded messages efficiently over wires. New York University art professor Samuel Morse began working on the concept in 1832. He coded letters as dots and dashes that could then be sent as electrical pulses to an electromagnetic receiver. The idea was not novel—by the time he presented his concept before the U.S. Congress in 1838, 62 people had claimed they had invented the first electrical telegraph—but Morse was the first to successfully arrange financial backing and launch his system.

Things weren't easy for Morse at first. He won little initial support for his idea from Congress and was forced to shop his ideas in Europe, where he was met with the same lack of interest. In Russia, Tsar Nicholas I refused the innovation; it was labeled an "instrument of subversion." In a letter to Francis O.J. Smith in 1838, Morse wrote: "This mode of instantaneous communication must inevitably become an instrument of immense power, to be wielded for good or for evil, as it shall be properly or improperly directed."

Entrepreneurs, Politicians, and Pirates: Congressional Support and Thievery

Morse continued to refine his system and raise support, finally meeting with success in 1842, when he strung wires between committee rooms of Congress and gave a successful live demonstration. In 1843, he was granted

$30,000 and the right to build a telegraph line from Washington, D.C., to Baltimore. On May 24, 1844, he transmitted the first message on that line: "What hath God wrought!"

The years that followed were difficult ones. In the rush to get in on the action, many telegraph entrepreneurs started, failed, and quit. Start-up expenses during the building of the telegraph network made the cost per message quite high, and that, coupled with a lack of compatibility in systems from region to region, dampened public enthusiasm. By the mid-1850s, most telegraph firms in the United States were bankrupt or nearly so. Fights over ownership rights landed in courtrooms. Pirates were copying Morse's patent and competing with him by building their own lines.

Despite all this, the telegraph was seen as a communications wonder. In the competitive market, built by self-funded pioneers with no government regulation, a 20-word message could be sent 500 miles in the United States for one dollar. In the United Kingdom, where telegraph development was slower and of a higher quality, the same message would cost $7 at a shorter distance.

A select group of powerful companies emerged, including Eastern Telegraph (later to become known as Cable and Wireless) in Britain and Western Union in the United States.

Structure and Regulation: International Agreement Makes the System Better

In France, Belgium, Russia, and other nations, the governments each controlled the telegraph business or allowed control to companies that would be responsive to government demands. So many governments with so many different systems made the world's networks incompatible. Between 1849 and 1855, most European nations had hammered out agreements to standardize the systems, and by 1865 they had agreed to adopt the Morse system and apparatus. The International Telegraph Union was founded to oversee rules of message priority, transmission, and delivery at a conference convened by Napoleon III of France. (The ITU still exists today as the International Telecommunications Union; it is now involved in helping set Internet policy.) In America, representatives of most of the major telegraph companies first met in 1857 to hammer out rules and standards. After years of chaos, common practices allowed for reliable connections, fair and predictable rates, and a common language, allowing the telegraph to mature as a medium.

The first consumers to accept the system, warts and all, were people whose business was dependent on information: newspapers, stockbrokers,

governments, and businesses. Personal messages were rarely sent; by 1851, social messages accounted for just 9 percent of telegraph traffic in the United
States.

Still, over the course of just two decades, the telegraph went from being
laughed at in the halls of Congress to being extolled as "a perpetual miracle."
Members of Congress had chuckled with doubt when Morse presented his prototype in 1838. When the first transatlantic cable was built from England to the
United States in 1858, a writer for the *Times* of London raved, "Since the discovery of Columbus, nothing has been done in any degree comparable to the
vast enlargement which has thus been given to the sphere of human activity."
Authors Charles F. Briggs and Augustus Maverick wrote in the 1858 study "The
Story of the Telegraph," "Of all the marvelous achievements of modern science
the electric telegraph is transcendentally the greatest and most serviceable to
mankind." And they added: "The whole earth will be belted with the electric
current, palpitating with human thoughts and emotions." These statements
were echoed in references to an updated wired technology more than 100 years
later, during the Internet's "awe" stage in the 1990s.

In 1864, Western Union operated on 44,000 miles of wire and was valued
at $10 million. Within the next year, its worth had jumped to $21 million. It is
estimated that between 1857 and 1867 the company's value grew by 11,000
percent. In 1866, its network included about 100,000 miles of wire and its capital stock value was in excess of $40 million.

By 1873, Western Union was carrying more than 12 million messages a year
and the telegraph was vital to communications across the world. Because it
was dominant, it began charging more for service than most of the other telegraph companies in the world; it was labeled a monopoly by many people.
Concern was expressed in Congress and elsewhere that Western Union's ability to "crush all rivals" would allow the company to "perform the least service
for the most money." Postmaster General John Creswell proposed that the
government take over the U.S. telegraph system. (This argument would be
echoed in the early 1990s, when some people spoke out for government regulation of the Internet in order to prevent its takeover by commercial interests.)

Many critics were concerned about the cozy relationship between Western
Union and the Associated Press—the news service that grew with and because
of the telegraph industry. AP was a loyal Western Union customer, accounting
for 12 percent of Western Union's total revenue by 1872. It was allowed pref-
erential rates and speediest transmission. It was said that AP rivals were de-

nied access to Western Union's service and that the AP reciprocated by failing to report bad news about Western Union. This alliance left the two companies as controlling powers in the flow of news in the United States.

It was also said that the telegraph was becoming a communications form reserved for the elite. Joseph Medill, publisher of the *Chicago Sun-Times,* wrote to Western Union president William Orton in 1872 in a letter reprinted in David Arnes Wells "The Relation of the Government to the Telegraph": "The telegraph . . . carries nothing more material than thought. The lightest of tolls should be charged for its labors, for it is one of the greatest educational instrumentalities of the nation and world. Its services should be as nearly free to the whole people as possible." (In the 1990s, the same sentiments were being expressed by many people who were concerned that the Internet would create a "digital divide" between rich and poor unless there was an intense effort to keep the new medium available to all. Their argument inspired the building of open Internet access points in public libraries and schools.)

At the end of the 19th century, demands for constraints on Western Union's power resulted in the passage of the Mann-Elkins Act of 1910, granting the Interstate Commerce Commission regulatory oversight of telegraph rates. Later, the Communications Act of 1934 switched regulation of the telegraph industry to the newly created Federal Communications Commission. By this time, the radio and telephone had diminished the impact of the telegraph.

Making the World a Little Smaller

Prior to the telegraph, communication in the 1830s was about the same as it had been in the years just after Gutenberg's invention of the printing press. It took days, weeks, and even months for messages to be sent from one location to a far-flung position. After the telegraph cable was stretched from coast to coast in the 1850s, a message from London to New York could be sent in mere minutes, and the world suddenly became much smaller.

Prior to the telegraph, politics and business were constrained by geography. The world was divided into isolated regions. There was limited knowledge of national or international news, and that which was shared was generally quite dated. After the telegraph, the world changed. It seemed as if information could flow like water.

By the 1850s, predictions about the impact of the new medium began to abound. The telegraph would alter business and politics. It would make the world smaller, erase national rivalries, and contribute to the establishment of world peace. It would make newspapers obsolete. All of the same statements were made in the 1990s by people who were wowed by the first-blush potential of the Internet.

RADIO

The Innovation: Persistence and Self-Promotion Pay Off for Marconi

Radio was the next long-distance communications medium to be developed by curious minds; a wireless wonder with much more going for it than the signal fires and drum codes of early humans. Like the flash of a fire and the thumping rhythm of a drum, the electromagnetic wave of the first radio signals could broadcast a message to anyone and everyone within a certain radius. Because of this, governments first feared its potential as a subversive technology. When Italian inventor Guglielmo Marconi brought an early model of his "wireless telegraph" to Britain in 1896, worried customs officials crushed the potentially dangerous device.

Since the 1820s, scientists had claimed that pulses of electricity could be carried invisibly over the air, and Scottish mathematician James Clerk Maxwell proved it logically in 1865. This moved curious thinkers of the time to work to capture the waves Maxwell had described, and in 1887 German scientist Heinrich Hertz did it. This spurred excitement about the potential for communicating without the telegraph's complicated wire network.

Marconi, a young man who had disappointed his prosperous family by failing to gain entrance to study at the University of Bologna, began tinkering with Hertz's ideas. In 1894, he used Hertz's coil wire and spark of electricity and he added a Branly coherer (a tube with electrical contacts on each end and metal dust in the middle), a battery, and a Morse printer. It worked from the start, but only over extreme distances. Marconi worked to refine all the pieces and was wirelessly sending Morse-code messages a kilometer away by 1895.

Entrepreneurs, Politicians, and Pirates: Marconi Uses His Wit, Connections

Other inventors in Russia and the United States at that time had been working on similar devices, but Marconi's commercial success resulted from his persistence and self-promotion and a little help from some friends. After failing to interest the Italian government in his wireless telegraph invention, Marconi, whose mother was Irish, traveled to Britain to sell the idea in 1896. His mother's cousin, Henry James Davis, helped repair the device after it was crushed by customs officials. He also networked with his friends, and within months Marconi was working on improvements on the device in the workshop of William Preece, chief engineer of the British Post Office. By early 1897, after many successful public demonstrations, Marconi, just 23 years old and now hailed as the "wizard of wireless," was awarded a British patent for his device.

Marconi gained financial backing and political connections thanks in great part to the publicity given to his work. He started the Wireless Telegraph and Signal Company. He saw that wireless was best used in ship-to-ship and ship-to-shore communications. The Italian navy and British government began using the machines, he supplied one to Queen Victoria for her personal use, and he promoted the device by filing minute-by-minute journalistic reports of the 1898 Kingstown Regatta and the 1899 America's Cup yacht race for newspapers. "We are learning to launch our winged words," reported the November 5, 1897, edition of the *New York Times*. Marconi went from being a disappointment to his parents to being a world-renowned inventor and entrepreneur in a span of just five years.

The first "pirates" of radio popped up when amateur wireless enthusiasts began experimenting, listening for signals and logging them—often on homemade wireless sets made of odds and ends ranging from bedsprings to Quaker Oats boxes. These "radioheads" were estimated to number up to 100,000 in the United States by 1912, and their signals sometimes interfered with important ship-to-ship and ship-to-shore radio traffic.

While Marconi was the innovator and first developer of radio, its great entrepreneur was David Sarnoff, a former Marconi employee who made the Radio Corporation of America into one of the most dominant communications companies in history. Sarnoff's role in the story of radio is directly tied to a move engineered by behind-the-scenes government-industry machinations in the 1920s.

Structure and Regulation: Marconi, "Radioheads," and RCA Inspire New Rules

At the turn from the 19th to the 20th century, Marconi's pricing practices began to anger some customers. Competing companies popped up all over the world; many of them pirating Marconi's technologies. By 1900, there were four competing wireless systems, yet Marconi's power was seen as a danger. In 1903, Germany's Kaiser Wilhelm put together an international conference on wireless. In 1906, participants in the Second International Radio Conference condemned the Marconi system's attempts to ignore Morse signals sent by other systems on ships at sea. In 1909, the year Marconi was awarded the Nobel Prize for physics, the British government bought all of his ship-to-shore transmitting stations in Britain for just $75,000. At the time, wireless was controlled in Britain by the post office, and in Germany wireless was controlled by a government-supported company, Telefunken. Most experimentation and investment in radio in the United States between 1902 and 1919 was accomplished by the U.S. Navy because of radio's importance in ship-to-ship and ship-to-land communications.

The extra traffic from the radioheads began to clog the airwaves, crossing or canceling out vital signals, including information about the sinking of the *Titanic* on April 14, 1912. Because of this, Congress passed the Radio Act of 1912, giving priority to distress calls and requiring that wireless operators be licensed. The act broke the radio spectrum into specific bands and required amateurs to broadcast within the short-wave piece of the spectrum. World War I proved the foibles of the telegraph: When wires were cut, the system went down. This reinforced the importance of the development of radio.

In the years just before World War I, scientists at companies such as American Telephone and Telegraph, General Electric, and Westinghouse, and inventors—including Reginald Fessenden, Lee De Forest, and Cyril Elwell—were mapping out ways they could develop the potential of wireless communication so it could broadcast more sophisticated messages than the dots and dashes of Morse Code.

By 1914, Fessenden, a Canadian who was once employed in Thomas Edison's labs, had worked with General Electric to build alternators that could sustain a consistent broadcast wave powerful enough to transmit voices and music over thousands of miles. Marconi saw the handwriting on the wall and offered to buy all of the alternators GE could produce. Officials in the U.S. Navy feared the potential for a firm headed up by a foreigner to hold a radio monopoly; they acted in 1919 to stop the deal by proposing that GE join in supporting the development of a new American radio company. At the time, the Navy possessed the patents, GE had the alternators and saw its patriotic duty, and the Radio Corporation of America (RCA) was formed in 1919 with the stipulation that no foreigner could hold more than 20 percent of the stock.

The success of RCA would hinge on a deal brokered by GE vice president Owen Young, who put together a patent-sharing arrangement that brought all key U.S. radio competitors together. Under the deal, RCA could use any radio patent held by the other firms. In exchange, companies received a percentage of RCA ownership. GE retained 30.1 percent of the stock; Westinghouse had 20.6 percent; AT&T owned 10.3 percent.

By 1923, RCA, led by former American Marconi Company employee David Sarnoff, built a dominant position, managing 50 percent of communications across the Pacific and 30 percent of the Atlantic traffic. A communications-hardware powerhouse had been built out of government-corporate coopera-tion. When AT&T was about to begin to market its own radio receivers and the Federal Trade Commission began a formal investigation of RCA's radio equipment monopoly, Sarnoff lobbied Washington officials with the message that radio needed a "superstructure" to keep it disciplined and orderly. In 1926 the broadcasting aspect of RCA was split from the equipment manufacturing when the National Broadcasting Company (NBC) was formed. It was owned 50 percent by RCA, 30 percent by GE, and 20 percent by Westinghouse. It paid $1 million for AT&T's New York station WEAF, the first in what would be-come the NBC radio network.

Europeans saw the U.S. commercial radio model as an undisciplined mess. The British Broadcasting Company, launched in 1923 by a group of radio manufacturers, was supported by a 10-shilling license fee paid to the British Post Office by all new radio owners. A ban on advertising, it was assumed, would create a public-spirited and unbiased system. The French evolved a public broadcast system in which the government operated a network of sta-tions and also allowed private stations that were tightly controlled by the gov-ernment. The Germans established a similar system, while governments in Italy, Czechoslovakia, and Austria allowed private monopolies that were basi-cally controlled by the state. Communist Russia's government controlled all radio broadcasting.

When airwave pirates began clogging the section of the radio spectrum as-signed to licensed broadcasters, the Radio Act of 1927 formally established U.S. government "control over all channels" and created the Federal Radio Com-mission. Over the next few years the FRC's decisions tended to support large commercial broadcasters, and NBC's network grew to 69 stations. RCA's stock price went from $85 in early 1928 to $500 by the summer of 1929. The stock market crash of 1929 dropped it down to $20 per share, but tough economic times of the 1930s couldn't stop the well-developed NBC network. NBC had some competitors, such as the CBS network, but in 1942, the FRC's successor, the Federal Communications Commission, required NBC to sell off one of its networks—it eventually became the American Broadcasting Company (ABC).

Making the World a Little Smaller

The 1920s were the decade in which radio boomed in the United States. People rushed to buy sets, and business and social structures adapted to the new medium. Universities began to offer radio-based courses; churches began broadcasting their services; newspapers created tie-ins with radio broadcasts.

By 1922 there were 576 licensed radio broadcasters and the publication *Radio Broadcast* was launched, breathlessly announcing that in the age of radio, "government will be a living thing to its citizens instead of an abstract and unseen force." As with television in later years, however, entertainment came to rule the radio waves much more than governmental or educational content, as commercial sponsors wanted the airtime they paid for to have large audiences. Most listeners enjoyed hearing their favorite music, variety programs that included comic routines and live bands, and serial comedies and dramas. Broadcasts of major sports events became popular as the medium matured and remote broadcasts became possible.

"Ham" or amateur radio operators—formerly known as radioheads—were individually licensed by the FCC to operate basement, bedroom, office, or kitchen stations through which they met people from around the world and also offered help in times of need.

Radio was a key lifeline of information for the masses in the years of World War II. Listeners around the world sat transfixed before their radio sets as vivid reports of battles, victories, and defeats were broadcast by reporters such as H.V. Kaltenborn and Edward R. Murrow. Franklin D. Roosevelt, Winston Churchill, Adolph Hitler, and other political leaders used the medium to influence public opinion.

THE TELEPHONE

The Innovation: Meucci Beat Bell, but Bell's Business Boomed

The idea behind the telephone—a device that would allow people to speak voice-to-voice with other individuals at a distance—was in existence long before Alexander Graham Bell was credited for bringing it to life in the United States in 1876. In fact, the word *telephone* came into use for such a device much earlier in the 19th century, and credit for its first invention is now officially attributed to Italian innovator Antonio Meucci.

Meucci was trained as a chemical and mechanical engineer in Florence. An alleged participant in the Italian Liberation Movement, he spent a short time in prison before marrying and emigrating to the United States via Havana,

Cuba. Meucci is said to have invented his first very basic "teletraphone" in Havana in 1849 to share information from one room to another in his house. He moved to the United States and developed as many as 30 different teletraphone models between 1850 and 1862. His phones, which used a paired electro-magneto transmitter and receiver, gave off a weaker signal than other such devices of the time, and could also expose users to burns.

Meucci's phone was demonstrated in the United States in 1860, and the event was described in an article in New York's Italian-language newspaper. It is said that Meucci was too poor to pay the $250 it would cost to apply for a patent. In 1874, he presented models of his device to a vice president of Western Union. He continually told his wife that one day his invention would make them rich. He was shocked to see Bell given credit for the telephone in 1876. Some action was taken to dispute Bell's patent, but Meucci's poor English skills and lack of business acumen were no match for Bell's connections. The issue was postponed year after year until Meucci died a poor man in 1889 and the case was dropped. More than 100 years later, an Italian group requested that Meucci's innovation be recognized, and in 2001 a U.S. congressional resolution officially credited Meucci with the invention of the telephone.

Other innovators of the time can be credited with work closely paralleling and/or contributing mechanical parts to the inventions of Bell and Meucci. Among them are: French telegrapher Charles Bourseul, who devised a phone in 1854; German scientist Johann Reis, who first transmitted musical tones— no voices or complex sounds—on his device in 1860; American Elisha Gray, founder of the Western Electric Company, who developed a vibrating transmitter in 1874; and the great American inventor Thomas Edison, whose carbon-grain transmitter invented at about this same time remained a standard part used in telephones until the 1980s.

Bell, a professor of vocal physiology at the University of Boston, began his research in 1874. He theorized that the vibrations of speech might be converted into magneto-induction currents that could be reproduced at a distance. On March 10, 1876, it is said that Bell spoke to his lab assistant, "Mr. Watson, come here, I want to see you," and Watson answered; Bell's electro-magnetic, sound-powered transmitter was completed. The first "long-distance" call came a few months later, when Bell "borrowed" telegraph lines to transmit a conversation between Paris, Ontario, and Brantford, Ontario, 10 miles apart.

When the first phone directory, or telephone "book," was issued in New Haven, Connecticut, in 1878, it was a single page containing just 50 listings and no phone "numbers." In the early days, phone calls were completed by manual line-switching performed by switchboard operators. The phone systems were so small at first that the operators simply identified the lines by family or business names. Assigned phone numbers were first introduced in Lowell, Massachusetts, in 1879, when a measles epidemic made it seem likely that the four switchboard operators who knew every one of the 200 phone subscribers' names might fall ill, leaving no one with knowledge to operate the system. There was a fear at the time that people would not like to be classified as numbers, but the assignation of numbers was practical and necessary, and it was thus adopted.

It took many more inventions over the years to make Bell's telephone an efficiently networked communications device. For instance, A.G. Smith and Almon Strowger came up with vital signal-switching devices. Strowger was a Kansas City mortician who claimed his business was being disrupted by local telephone operators who were steering customers elsewhere. His 1891 invention of the Strowger Automatic Telephone Exchange enabled the removal of the switchboard operator at certain levels of the system.

Entrepreneurs, Politicians, and Pirates: Western Union Made a Big Mistake

While Bell may not be the first inventor of the telephone, he popularized the device and was behind the development of the first public-switched telephone network. His work was backed by a number of financiers, including Gardiner G. Hubbard and Thomas Sanders, who had formed the Bell Patent Association in 1875. The men both had daughters who had hearing difficulties, and they had engaged Bell's services as a speech pathologist. They came to know him and his invention, and became invested in it.

The inventor and his investment group struggled at first while trying to promote their new device; at one point in 1876 they offered to sell their patents to Western Union Telegraph Company for $100,000. The response formed by a committee appointed to investigate the offer is reported in various locations, including in the FCC's "Condensed History of Telecommunications." It reads, in part:

> We do not see that this device will be ever capable of sending recognizable speech over a distance of several miles. Messer Hubbard and Bell want to install one of their "telephone devices" in every city. The idea is idiotic on the face of it. Furthermore, why would any person want to use this ungainly and impracti-

cal device when he can send a messenger to the telegraph office and have a clear written message sent to any large city in the United States? . . . Mr. G.G. Hubbard's fanciful predictions, while they sound rosy, are based on wild-eyed imagination and lack of understanding of the technical and economic facts of the situation, and a posture of ignoring the obvious limitations of his device, which is hardly more than a toy. . . . This device is inherently of no use to us. We do not recommend its purchase.

Years later, Bell's "idiotic" phone company bought out Western Union.

Bell's first U.S. telephone exchange began operations in 1878, and rivals began to pop up. Western Union established several phone subsidiaries, and in addition to defending the challenge of his patent from Meucci, Bell was also tied up in court while protecting his patent from infringements by Edison and Gray. In 1879, Bell merged the New England Telephone Company and the Bell Telephone Company, forming the National Bell Telephone Company in an effort to speed the adoption of the telephone. In 1880, Bell merged this company with others to form the American Bell Telephone Company. Bell Laboratories got its start at about this time, first formed as a unit called the Electrical and Patent Department. In 1885 American Telegraph and Telephone Company (AT&T) was formed; it dominated telephone communications for the next century. At one point in time, Bell System people purposely denigrated the U.S. telephone system to drive down stock prices of all phone companies and thus make it easier for Bell to acquire smaller competitors. In the case of the telephone, the system initiator was also the pirate force.

By 1900 there were already nearly 600,000 phones in Bell's telephone system; that number shot up to 2.2 million phones by 1905, and 5.8 million by 1910. In 1915 the transcontinental telephone line began operating; in honor of the occasion, Alexander Bell got on a line in New York and spoke to Thomas Watson in San Francisco, saying, "Mr. Watson, come here; I want you."

Structure and Regulation: "Ma Bell" Monopoly Broken by Divestiture

The expiration of Bell's first telephone patent in 1894 allowed independent phone systems to spring up in many cities—as many as 6,000 of them by 1897. Because Bell refused to share facilities with other companies, the public in these areas had to subscribe to both the AT&T service and their independent service in order to be able to send and receive long-distance phone calls.

By 1907, AT&T had a near monopoly on phone and telegraph service, thanks to its purchase of Western Union. Its president, Theodore Vail, urged at the time that a monopoly could most efficiently operate the nation's far-flung communications network. At the urging of the public and AT&T competitors, the government began to investigate the company for antitrust violations, thus forcing the 1913 Kingsbury Commitment, an agreement between AT&T vice president Nathan Kingsbury and the office of the U.S. attorney general. Under it, AT&T agreed to divest itself of Western Union and provide long-distance services to independent phone exchanges. The agreement was not effective in stopping AT&T from solidifying its monopoly by taking over all of the most profitable urban markets; it strengthened AT&T's dominance.

During World War I, the government nationalized telephone and telegraph lines in the United States from June 1918 to July 1919, when, after a joint resolution of Congress, President Wilson issued an order putting them under the direction of the U.S. Post Office. A year later, the systems were returned to private ownership, AT&T resumed its monopolistic hold, and by 1934 the government again acted, this time agreeing to allow it to operate as a "regulated monopoly" under the jurisdiction of the FCC.

Public utility commissions in state and local jurisdictions were appointed regulators of AT&T and the nation's independent phone companies, while the FCC regulated long-distance services conducted across state lines. They set the rates the phone companies could charge and determined what services and equipment each could offer.

This stayed in effect until AT&T's forced divestiture in 1984, the conclusion of a U.S. Department of Justice antitrust suit that had been filed in 1974. The all-powerful company had become popularly known and disparaged as "Ma Bell." AT&T's local operations were divided into seven independent Regional Bell Operating Companies, known as the "Baby Bells." AT&T became a long-distance-services company.

Making the World a Little Smaller

Within 50 years of its invention, the telephone had become an indispensable tool in the United States. By 1948, the 30 millionth phone was connected in the United States; by the 1960s, there were more than 80 million phone hookups in the United States and 160 million in the world; by 1980, there were more than 175 million telephone subscriber lines in the United States. In

1993, the first digital cellular network went online in Orlando, Florida; by 1995 there were 25 million cellular phone subscribers, and that number exploded at the turn of the century, with digital cellular phone service expected to replace land-line phones for most U.S. customers by as early as 2010.

After the phone's invention in the late 19th century, people raved about its positive aspects and ranted about what they anticipated would be negatives. Their key points, recorded by Ithiel de Sola Pool in his 1983 book *Forecasting the Telephone,* mirror nearly precisely what was later predicted about the impact of the Internet. For example, people said the telephone would: help further democracy; be a tool for grassroots organizers; lead to additional advances in networked communications; allow social decentralization, resulting in a movement out of cities and more flexible work arrangements; change marketing and politics; alter the ways in which wars are fought; cause the postal service to lose business; open up new job opportunities; allow more public feedback; make the world smaller, increasing contact between peoples of all nations and thus fostering world peace; increase crime and aid criminals; be an aid for physicians, police, fire, and emergency workers; be a valuable tool for journalists; bring people closer together, decreasing loneliness and building new communities; inspire a decline in the art of writing; have an impact on language patterns and introduce new words; and someday lead to an advanced form of the transmission of intelligence.

Privacy was also a major concern. As is the case with the Internet, the telephone worked to improve privacy while simultaneously leaving people open to invasions of their privacy. In the beginning days of the telephone, people would often have to journey to the local general store or some other central point to be able to make and receive calls. Most homes weren't wired together, and eavesdroppers could hear you conduct your personal business as you used a public phone. Switchboard operators who connected the calls would also regularly invade people's privacy. The early house-to-house phone systems were often "party lines" on which a number of families would receive calls, and others were free to listen in and often chose to do so.

Today, while most homes are wired and people can travel freely, conducting their phone conversations wirelessly, wiretapping and other surveillance methods can be utilized to listen in on their private business. People's privacy can also be interrupted by unwanted phone calls from telemarketers and others who wish to profit in some way—just as Internet e-mail accounts receive unwanted sales pitches, known as "spam." The invention of the telephone also worked to increase privacy in many ways. It permitted people to exchange information without having to put it in writing, and a call on the phone came to replace such intrusions on domestic seclusion as unexpected visits from relatives or neighbors and the pushy patter of door-to-door salesmen. The same could be said for the Internet—privacy has been enhanced in some ways because e-mail and instant messaging have reduced the frequency of the jangling interruptions previously dished out by our telephones.

TELEVISION

Early Predictions: The Coming of the "Telectroscope"

Writers such as Walter Scott, Jules Verne, Mark Twain, and H.G. Wells began postulating the idea of "seeing at a distance"—as the earliest concepts of television were predicted—in the 18th and 19th centuries. A set of Victorian trading cards labeled "100 Years Hence" depicted future technological developments including one titled "Concerts and Opera at Home." The card features a funnel-shaped device projecting the image of a live concert on a wall—the device is similar to the projection televisions introduced in the 1990s.

As with radio, scientists from around the world worked to perfect television for decades, with the key breakthroughs coming in the early 20th century, the invention springing from the work of many minds. In the early 1870s, the discovery of the photosensitive properties of selenium led researchers to posit the possibilities of seeing remotely through the use of electricity. Between 1878 and 1880, theoretical studies outlining the principles of television were published by Portuguese professor Adriano de Paiva and articles began popping up in publications around the world (including *Scientific American*, *Harper's*, and the *Times* of London) about the idea of an "electric telescope" or "telectroscope." Alexander Graham Bell got into the act, writing many articles and research papers.

The 1900 Paris Exposition looked to the future. There, at the first International Congress of Electricity, Russian Constantin Perskyi read a paper in which he described a mechanical method of receiving live images and sound, calling the device "television." This is said to be the first use of the term.

The Innovation: From Mechanical Television to Farnsworth's Image Dissector

The medium would not be invented without the cathode-ray tube (CRT). German physicist Eugen Goldstein was the first scientist to label the cathode ray, and in 1876 he showed how these rays could cast shadows and be deflected by magnetic fields. Experiments with cathode rays also led to the discovery of X-rays and the electron. In television sets and computers, a CRT is a tube containing a cathode and a luminescent screen; each time an electron beamed from the cathode hits the screen, a spot of light appears; a combination of such lines of light forms a picture. CRTs later became a key component in personal computers.

The next important invention came in 1884. Young German engineering student Paul Gottlieb Nipkow invented the scanning disk that became the key element in the mechanical television systems developed over the next 50 years. The Nipkow disk dissected images, allowing them to be transmitted sequentially over wire, allowing inventive thinkers to prove that pictures could be transferred electrically. American inventor Charles Francis Jenkins transmitted pictures of President Harding from Washington to Philadelphia by radio in 1923, and he demonstrated a mechanical television scanning system using a revolving disk in 1925. Television was not the term popularly adopted at the time, and Jenkins called his invention "radiovision." He correctly predicted in a 1925 article titled "Home Radio Movies": "It will not be very long now before one may see on a small white screen in one's home notable current events, like inaugural ceremonies, ball games, pageants, as well as pantomime performance broadcast from motion-picture film."

In the late 1920s and early 1930s, Jenkins's motor-driven mechanical scanning television system was demonstrated a great deal in the United States, and it was his system by which many people in New York witnessed television for the first time. The images were made up of 40 to 48 lines on a six-inch-square mirror. Because the bulky system used the slow process of mechanical scanning, the images were too murky to be seen with any clarity, and announcers had to fill in most of the visual details.

American Telephone & Telegraph also got into the act in the early days of television, transmitting moving images of Herbert Hoover from Washington to New York over phone circuits in 1927 using a 185-line system developed by Herbert E. Ives. It was front-page news in the April 28 edition of the *New York Times*. Dr. C.G. Abbot, secretary of the Smithsonian Institution, said at the time: "Perhaps 50 years hence there may be other developments, needing no apparatus whatever, by means of which human thought may be transmitted at the will of the sender by the process which we now dimly recognize as telepathy."

In 1928, General Electric began broadcasting a 24-line mechanical image from a New York station thanks to engineer Ernest Alexanderson's development of a mechanical television system. That same year, German Denes von Mihaly appeared at the Berlin Radio Show and demonstrated his 30-line mechanical television system. Also in the 1920s, Kenjiro Takayanagi was making strides in "wireless distance vision" in Japan and Scottish engineer John Logie Baird was building a working mechanical television system in Britain. All of these inventive men accomplished firsts, but none of them is seen as the "inventor" of TV.

Philo Taylor Farnsworth developed what he called the "image dissector," the first working electronic camera tube, in San Francisco in 1927. As a youngster growing up in Utah and Idaho he'd read in a magazine about the idea of the broadcasting of images and sound, becoming so fascinated that he was motivated to study molecular theory and electricity. By the age of 14 this daydreaming farm boy had hatched the idea that would lead to his eventual breakthrough to the development of television: You could utilize an electron beam as a scanning device that could send a picture, just as your eyes scan a page to send visuals to your brain.

Entrepreneurs, Politicians, and Pirates: TV Inventors and Backers Battle in Court

Farnsworth worked as a radio repairman and for a railroad yard before he convinced backers to finance his television research at Crocker Research Laboratories in California in 1926 to "take all of moving parts out of television." Farnsworth was just 20 years old at the time. By 1927, with the help of his wife Pem and her brother Cliff Gardner, he had transmitted the images of a straight line, a puff of smoke, and a dollar sign (because an investor had asked him, "When are we going to see some dollars in this thing, Farnsworth?"). Backer George Everson sent a telegram to another investor; it read: "The damned thing works!" Farnsworth Radio and Television Corporation was founded in 1938, and the inventor's patents covered scanning, contrast, controls, synchronizing, focusing, and power.

By the late 1920s, Radio Corporation of America president David Sarnoff was intrigued with Farnsworth's work, and he sent engineer Vladimir Zworykin to Farnsworth's lab to see what the brash kid was doing. Zworykin visited for three days. He returned to RCA and by 1933 he had perfected his "inconoscope"—an invention nearly identical to Farnsworth's image dissector. A patent battle ensued.

Zworykin, an engineering Ph.D., had previously been in on one of the earliest television research projects when he worked at the Imperial Institute of Technology in Russia with Boris Rosing. Zworykin was hired to be the director of RCA's Electronic Research Laboratory in 1929, after demonstrating a television receiver that included what he called a "Kinescope" tube. Fellow Russian immigrant Sarnoff saw its value and hired Zworykin to work with RCA Victor in Camden, N.J.

Zworykin had applied for a television patent for his first practical conception of a camera tube in 1923, but during its 1930s TV patent battle with Farnsworth, RCA could not produce any evidence that Zworykin had produced an operable television system in the 1920s. Farnsworth was able to prove his origination of the idea when his high school science teacher, Justin Tolman, came forward with an original sketch of an electronic tube—the image dissector—that Farnsworth had drawn for him in 1922. Thanks to Tolman, Farnsworth was awarded priority of invention by the U.S. Patent Office in 1934. Litigation followed for many years, with a series of appeals before Sarnoff finally agreed to pay Farnsworth royalties.

Philo Farnsworth had won this long-fought 1930s TV patent battle but lost the war thanks to a war. Television broadcasting was still in its experimental stages but ready to take off when the U.S. government suspended sales of television sets and FCC licensing of commercial broadcasters during World War II. By the time televisions were again cleared for production, Farnsworth's patents had neared their expiration. RCA stepped in as soon as it could, becoming a dominant force in production and sales of television sets, promoting Zworykin and Sarnoff nationally as the "fathers of television."

Farnsworth suffered from such deep depression that he underwent shock treatments. He became an alcoholic and his house in Maine burned. A 2003 *Time* magazine essay written by Neil Postman reported that the inventor's son Kent said Farnsworth came to despise his invention: "I suppose you could say that he felt he had created kind of a monster, a way for people to waste a lot of their lives. . . . Throughout my childhood his reaction to television was, 'There's nothing on it worthwhile, and we're not going to watch it in this household, and I don't want it in your intellectual diet.'" Farnsworth didn't win universal respect for his television innovations until after his death in 1971.

Structure and Regulation: Manufacturers and Government Take Years to Agree

From the mid to late 1930s, a number of experimental television broadcast stations appeared across the United States, and radio behemoths NBC and CBS spent millions to investigate the new medium, building New York stations. In England, Baird had switched his system in 1936 to use of the Farnsworth image dissector, while EMI Marconi used Zworykin's iconoscope. Farnsworth's system was used in Germany to broadcast the 1936 Olympics. That same year, the Radio Manufacturer's Association (today known as the Electronics Industry Association) established a committee to hammer out standards for commercial television in the United States.

Over the objections of a majority of the RMA membership, RCA's 343-line/30-frame system was adopted by the committee. The RCA system was not accepted by other companies in the industry because of this disagreement, and commercial development was stalled. A 1940 FCC hearing again ended with no conclusive guidelines after manufacturers DuMont, Zenith, and Philco argued with each other and RCA. In June of 1941, the FCC finally announced the 525-line/30-frame, black-and-white system that is still in existence, but full-scale commercial broadcasting did not begin in the United States until 1947.

As in the case of radio, the United States allowed free-market commercial development of the new medium while most other nations chose to retain government control over most or all television broadcasters. While it's true that this led to the broadcast of a great deal of entertainment-oriented dreck, it has also led to more solid television content and freedom than can be found in any international broadcast system.

Making the World a Little Smaller

In its earliest years, most television programming was difficult to see. The CRTs were only a few inches wide, the picture was in black-and-white, and there were many parts of the country in which viewers could not receive a strong signal. Initial programming tended to be rather stagnant—many broadcasts were based on radio shows, leading more than one critic to complain that there was no reason to see the picture when nothing was going on.

Television replaced radio as the dominant broadcast medium by the 1950s and took over home entertainment. Approximately 8,000 U.S. households had television sets in 1946; 45.7 million had them by 1960. The popular programs of the 1950s and '60s boom days were serial dramas and comedies and variety shows that featured popular entertainers. Sports broadcasts were a big draw, and many families purchased televisions so they could follow their favorite teams.

The pace of innovation and improvements in television and other technologies developed in the United States over the past 100 years has been rapid, thanks to a confluence of several factors: the competitive atmosphere of the free-market economy; the laid-back role of government, which is a watchdog thus far preventing wholesale monopolies; and the spirit of invention and entrepreneurship prevalent in U.S. society.

The worldwide success of the freewheeling U.S. film and television industries over the course of the 20th century has spread images of the American culture—good and bad—to the most distant corners of the planet. It has made entertainment one of the nation's most lucrative and influential exports.

THE INTERNET

Cold-War Fears Generate a New Communications Technology

The communications technologies developed over the past 250 years have progressed in refinement, complexity, and social, political, and economic impact. While the innovations of the telegraph, radio, the telephone, and television followed quickly on one another's heels, the public Internet came along after four decades of television dominance and decades of private Internet use and development. It came along after hundreds of years of inventive thinking and groundbreaking theorizing, and it built on every bit of human intelligence that had come before.

The Internet is an invention that can never be pinned down as being due to one key innovator, one key invention, nor one key entrepreneur. Thus, this section is not subdivided in the same fashion as those on the telegraph, radio, and television. The key innovators were dozens of scientists whose work covers decades; the entrepreneurs were thousands of political leaders, policy wonks, technology administrators, government and commercial contractors, and even grassroots organizers; the pirates have taken various forms in a highly competitive atmosphere that includes hackers and self-described cryptoanarchists; and regulation is continually evolving because the system itself is in a near-constant state of massive evolution.

As is often the case in humankind's inventiveness, the fear of and preparation for war was the original driving force behind both the computer and networked digital communications. Following the devastating world conflicts of the first half of the 20th century and the introduction of nuclear weapons, the stakes were high; the governments of the world's superpowers began pouring monetary resources and big brainpower into gaining a strategic edge.

First Steps: Computers Came Before Networked Computing

Before the development of the first computer network came the developments leading up to the invention of computers. Early peoples used the abaci and abacus as tools to count and calculate. The algorithm was a 12th-century innovation, and in the 17th century John Napier invented logarithms and developed machines for multiplication. In 1642, Blaise Pascal created an adding machine that could "carry" numbers.

In 1833, Charles Babbage's idea for the "analytical engine" possessed the key components of the modern-day computer—it wasn't actually built successfully until the 1930s. And in 1844, George Boole described his Boolean system of logical and symbolic reasoning vital to today's computing. Vannevar Bush's invention of the differential analyzer in 1925 brought the ability for integration and differentiation in electrical computation. Up to then, most of the development had centered on numerical computations. In the 1930s, mathematician Alan Turing proposed the design of a general-purpose computer. Building on this, Claude Shannon demonstrated in the 1940s that Boolean algebra and logic could work in the mating of switching and computer circuits.

During and after World War II, U.S. Army and Navy contracts financed the development of computers. Researchers at the University of Pennsylvania built the first all-purpose, electronic digital computer, ENIAC (Electronic Numerical Integrator and Calculator), in 1946. It had about 18,000 vacuum tubes, filled a large room, and consumed as much electrical power as a small town. The first computer to operate on software was built a year later by engineers at IBM led by Thomas Watson. The 120-foot-long, million-dollar machine had 12,500 vacuum tubes and 21,400 mechanical relays. That same year, 1947, John Bardeen, Walter Brattain, and William Shockley invented the transistor, allowing computers (and radios and TVs) to operate without huge vacuum tubes.

Competing researchers Jack Kilby of Texas Instruments and Robert Noyce and Gordon Moore of Fairchild Semiconductor (the Farnsworth and Zworykin of the IC) developed the integrated circuit in 1958. By the early 1970s, researchers at IBM and Intel fit more components on a single chip, calling it the microprocessor, and computers became smaller as more computing potential was integrated into smaller and smaller chips.

In 1965, Moore, a co-founder of Intel, noticed that the number of transistors per chip was doubling and the circuits' performance increasing by 35 percent every 18 months. This became known as "Moore's Law." It is respected because it continues to hold true, years into the 21st century.

According to researcher Daniel Bois, between 1947 and 1997, microelectronics were miniaturized at a rate of 15 percent yearly, with an annual cost reduction of 30 percent, and a 50 percent improvement in function. The economies of scale involved in the miniaturization of computing power led to the development of personal computers and the opportunity for everyone—not just scientists and well-heeled corporations—to become a part of the coming Internet communications revolution.

Networks Developed through Trial and Error

President Dwight D. Eisenhower initiated the Advanced Research Projects Agency (ARPA) after the Soviet launch of Sputnik I in 1957. Eisenhower, a U.S. general during World War II, believed in the great value of science and formed ARPA in a quest for "the scientific improvement" of U.S. defense. This project employed, at one time or another, some of the finest engineers and research scientists of the late 20th century. The organization's early emphasis was on missile-defense systems and the detection of nuclear bomb tests. It wasn't until 1962, when J.C.R. Licklider arrived on the management team, that ARPA began investigating the idea of networked computers.

It's difficult for people who were born after 1985 to understand the WBI—the world before the Internet. Previous to the 1970s and 1980s, computers were extremely expensive, most were so large they could fill an entire room, and they had one keyboard at one workstation that was not networked to any other computer anywhere. Because the "sophisticated" computers of that time (which held less power than any PC you could buy today) were expensive to own and operate, many organizations "shared" time on them or bought computing time from another organization that owned a computer. These computers were lunky conglomerations of metal, plastic, and cables, isolated in their uses and content, and usually individually programmed to perform specific tasks.

In the early 1960s, Licklider, Leonard Kleinrock, Paul Baran, Lawrence Roberts, and other research scientists came up with the ideas that allowed them to individually dream of and eventually come together to create a globally interconnected set of computers through which everyone could quickly and easily access data and programs from any site. Licklider—known by one and all as "Lick"—jokingly called it an "Intergalactic Computer Network," but he and his team began to seriously build the ideas and the technology that turned out to be the Internet. By 1964, some researchers had begun using their enormous mainframe computers to occasionally trade information by an early, informal form of e-mail—but the purpose wasn't to formulate a research network; they were just trying to get their work done efficiently.

Earlier in the '60s, Baran and British scientist Donald Davies had individually proposed the idea of sending blocks of data—packets—through a digital network (just as Farnsworth and Zworykin had individually developed the idea of the television tube). Packet-switching could be seen as the exchanging of digital "envelopes" with a receiver's address and a sender's address. Roberts and many others got down to the serious business of taking this concept, combining it with other researchers' proposals, and building a real network in 1967, 1968, and 1969.

In 1968, Licklider teamed with Robert Taylor to write a paper titled "The Computer as a Communications Device." In it, they defined four principles by which computers could making lasting contributions: (1) Communication is defined as an interactive creative process; (2) Response times must be short to make conversations free and easy; (3) Larger networks can be formed from smaller regional networks; (4) Communities will form out of affinity and common interests.

"First, life will be happier for the on-line individual," they wrote, "the people with whom one interacts most strongly will be selected more by commonality of interests and goals than by accidents of proximity. Second, communication will be more effective and productive, and therefore more enjoyable. Third, much communication and interaction will be with programs and programming models . . . both challenging and rewarding. And fourth, there will be plenty of opportunity for everyone (who can afford a console) to find his calling, for the whole world of information, with all its fields and disciplines will be open to him." They predicted, "In a few years, men will be able to communicate more effectively through a machine than face to face."

Significant additional developments popped up over the next 25 years.

1969 Brought the First Request for Comments and Arpanet

Critical work on the first real network was being completed, and information had to be shared between far-flung research groups. Steve Crocker, a young computer scientist, wrote a long memo—the first of what came to be called a Request for Comments (RFCs). These are, to this day, the accepted way in which computer networking engineers and scientists from around the world suggest, review, and adopt new technical standards. They also use RFCs to occasionally share networking humor. Since the day Crocker wrote the first Request for Comments, thousands more have followed. RFCs are a rich source of history about the development of the Internet, and anyone who has an interest in them can read them all online.

The researchers also established a name for themselves at this time—the Network Working Group. The democratic way in which decisions were made by these pioneers became a basis for the free-speech, free-exchange format of the Internet. The first network, ARPANET, went online in an extremely basic way in late 1969, connecting four major universities: the University of California at Los Angeles, Stanford (University) Research Institute, the University of California at Santa Barbara, and the University of Utah.

This rough system gave computer scientists and engineers the opportunity to begin refining ideas for a more efficient, reliable communications network. They had a lot of work to do in the years to come to get the "bugs" out, brainstorming, trying and failing, exchanging RFCs, and improving the system.

1970–1975 Saw Further Development of the Network and a Virtual Community

In 1970, ARPANET machines 5 through 12 were operating at locations around the country, including those at a Network Control Center at the technology corporation Bolt Beranek & Newman, at Harvard University, the RAND Corporation, and the Massachusetts Institute of Technology. File-Transfer Protocol (FTP)—the method for allowing computers to exchange files—was posted as RFC 354 in July 1972. ARPANET was unveiled to an audience for the first time at the International Conference on Computer Communication in October 1972.

The first electronic mail delivery engaging two machines was accomplished in 1972 by Ray Tomlinson—also the originator of the use of the @ symbol to indicate an e-mail address. By 1973, three-quarters of all traffic on the network was e-mail—still mostly researchers sharing information. An e-mail list group of the time named MsgGroup is believed to have been the first "virtual community."

The scientists had been using Network Control Protocol (NCP) to transfer data from one computer to another running on the same network. Vinton Cerf of UCLA and Stanford and Robert Kahn from ARPA came up with the ideas for Transmission Control Protocol and Internet Protocol (TCP/IP) over a span during the mid- to late-1970s. The superior TCP/IP allowed diverse computer networks to interconnect and communicate with one another no matter what network they were on at the time of use. TCP/IP made the Internet faster and more efficient; it was thus possible to bring more computers online at a lower price. This fueled the growth of the Internet.

1976–1979 Developments Include Ethernet and Emoticons

In 1976, Robert Metcalfe developed Ethernet, which allowed data to be transferred at rapid speeds over coaxial cables. Soon after, a packet satellite project (SATNET) that used satellites to link the United States with Europe was completed, thus a basic worldwide data-delivery universe was born.

Jimmy Carter's presidential campaign staff sent out e-mail several times a day in the fall of 1976, earning him the descriptor "computer-driven candidate." In 1979, Kevin MacKenzie, a member of the MsgGroup e-mail list, complained about the "loss of meaning," the lack of facial expressions, vocal inflection, and gestures in e-mail correspondence. He suggested the use of a new form of punctuation in e-mails and used the example -). This was far less sophisticated than the :o) and many other emoticons in use today. MacKenzie was flamed (criticized) by the other people in the e-mail group at the time, but his legacy lives on.

At this stage in its development, few people outside the research community used the Internet.

1980–1989 Network Expansion Inspires the IAB, IETF, and IRTF

The National Science Foundation started the Computer Science Research Network (CSNET) and had more than 70 sites online by 1983. In the mid-1980s, a coordinating group called the Internet Activities Board centralized networking efforts; late in the decade its membership numbered in the hundreds, and it was split into two groups, the Internet Engineering Task Force (IETF) and the Internet Research Task Force (IRTF).

In 1985, NSFNET, a "backbone" to connect five supercomputer centers located all over the United States, allowed the establishment of regional networks around the country, making a brighter, more connected future possible for more people. By 1986, most U.S. computer science departments were connected through this method, paying the NSF annual operation fees in order to use the network. More, mostly researcher-oriented networks emerged, including BITNET, USENET, and UUCP.

In 1989, Tim Berners-Lee, the researcher at CERN, the particle physics laboratory near Geneva, wrote the memo to his supervisors suggesting his ideas for the invention of a worldwide network that would revolutionize everything.

Making the World a Little Smaller

After Berners-Lee brought his "World-Wide Web" to life in 1990, and Andreessen launched Mosaic, the revolutionary browser, in 1993, the Internet had an estimated 16 million users by 1995, and venture capitalists were busy full-time, funding hundreds of new Internet-related business concerns.

Thanks to the work of thousands of collaborators over the final four decades of the 20th century, today's Internet is a continually expanding worldwide network of computer networks for the transport of myriad types of data. In addition to the names above, there were direct contributions from Ivan Sutherland, Robert Taylor, Alex McKenzie, Frank Heart, Jon Postel, Eric Bina, Robert Cailliau, Tom Jennings, Mark Horton, Bill Joy, Douglas Engelbart, Bill Atkinson, Ted Nelson, Linus Torvalds, Richard Stallman, Dave Clark, and so many others—some of them anonymous hackers or users—it is impossible to list them all here.

In 1996, there were approximately 45 million people using the Internet. By 1999, the number of worldwide Internet users reached 150 million, and more than half of them were from the United States. In 2000, there were 407 million users worldwide. By 2004, there were between 600 and 800 million users (counting has become more and more inexact as the network has grown, and estimates vary).

The Internet has been adopted as a method of expression and communications by people all over the globe. Wireless satellite and broadband communications networks are helping people in even the most remote locations find ways to connect. Overcoming the initial concerns that commercialization would limit creativity or freedom of speech, the Internet has become a crazy-quilt mix of commercial sites, government information, and incredibly interesting pages built by individuals who want to share their insights. The number of people making Internet pages continues to grow. As of 2005, more than 66.3 million domain names had been registered, approximately one for every 100 people living in the world.

3

Web Gems

Social, political, and economic expectations inspired intriguing statements about the Internet

Between 1990 and 1995, Internet stakeholders waged conflicts over the new technology. Engineers and computer scientists were struggling to find solutions for networking issues such as bandwidth and switching and teaming up to identify universal network protocols; politicians were voicing concern over improper materials (pornography, hatemongering) being promulgated online; law and national-security officials were advocating government-sanctioned encryption methods to fight crime and terrorism; entrepreneurs were trying to stake out as much valuable territory as possible; and socially conscious citizens were working to see that the new technology would be used for the greater good of all people.

It was a time when everyone had something to say, and because all of the voices could be heard, what everyone had to say made a difference. For the first time in history, a great majority of the battles tied to a technology—for ideals, for freedom, for power, for money—were fought on a fairly level playing field: The conflict over the Internet has been waged *on* the Internet. In its creation and unfolding over the past few decades, thousands of people played vital roles in giving it form, and it continues to evolve as a collaborative work. Writer Douglas Coupland commented in a 1994 article in *Wired* magazine: "There's a big cinder block stuck on the technology accelerator pedal, and we're only gonna go faster and faster, never stopping."

Leaders of the early 1990s knew that the decisions made in the developmental stages of the Internet would come to change people's sense of self, of space, of community, and of relationships—even down to the molecular level. Communications theorist Marshall McLuhan's global village, in which "centers are everywhere and margins are nowhere," had come to pass. This new technology would not only reshape our social spaces; its ability for embeddedness would also come to consume our social spaces.

In 1990, Mitchell Kapor and John Perry Barlow founded the nonprofit Electronic Frontier Foundation to address political issues surrounding the Internet. In their founding statement, they wrote:

> What is free speech, and what is merely data? What is a free press without paper and ink? What is a "place" in the world without tangible dimensions? How does one protect property which has no physical form and can be infinitely and easily reproduced? Can the history of one's personal business affairs properly belong to someone else? Can anyone morally claim to own knowledge itself? These are just a few of the questions for which neither law nor custom can provide concrete answers. In their absence, law-enforcement agencies such as the Secret Service and FBI, acting at the disposal of large information corporations, are seeking to create legal precedents which would radically limit Constitutional application to digital media. [It] threatens to become a long, difficult, and philosophically obscure struggle between institutional control and individual liberty.

Technology experts and theorists Esther Dyson, George Gilder, Jay Keyworth, and Alvin Toffler wrote a 1994 article titled "Magna Carta for the Information Age" for *New Perspectives Quarterly*. In it, they said, "The central event of the 20th century is the overthrow of matter. . . . The powers of mind are everywhere ascendant over the brute force of things. As humankind explores this new electronic frontier of knowledge, it must confront again the profound questions of how to organize itself for the common good. The meaning of freedom, structures of self-government, definition of property, nature of competition, conditions for cooperation, sense of community and nature of progress will each be redefined for the Knowledge Age—just as they were redefined for a new age of industry some 250 years ago."

Futurist Jim Dator looked far into the distance at the 1991 conference of the World Futures Studies Federation, saying, "In the early 21st century, the electronic 'information society' will be replaced by societies based on genetic and molecular engineering. . . . This portends forms and processes of 'participation,' and 'democracy' that are presently beyond my ability to imagine in sufficient detail."

In his 1994 book *Cyberia: Life in the Trenches of Hyperspace* Douglas Rushkoff wrote, "Coping in Cyberia means using our currently limited human language, bodies, emotions, and social realities to usher in something that's supposed to be free of those limitations. . . . Cyberia is frightening to everyone. Not just to technophobes, rich businessmen, Midwestern farmers, and suburban housewives, but, most of all, to the boys and girls hoping to ride the crest of the informational wave. Surf's up."

Fellow author William Gibson observed in a 1995 *Maclean's* magazine interview: "We are being shoved up against futurity with such violence that science fiction may become a historical term. . . . The Internet may be important because we are seeing something akin to what we did when we invented cities."

IDEAS REGARDING PROPERTY UNDERGOING CHANGE

The social commentators of the early 1990s told us that the current concept of ownership of property is under threat of obsolescence. Electronic Frontier Foundation co-founder John Perry Barlow wrote in the 1994 *Wired* magazine essay "The Economy of Ideas": "We're going to have to look at information as though we'd never seen the stuff before. . . . The economy of the future will be based on relationship rather than possession. It will be continuous rather than sequential. And finally, in the years to come, most human exchange will be virtual rather than physical, consisting not of stuff but the stuff of which dreams are made. Our future business will be conducted in a world made more of verbs than nouns." And *Wired* magazine editor Kevin Kelly added in a London Guardian interview titled "In 2004 We'll All Live on the Internet with Silicon Valley Visionaries," "Every effort to restrict copying is doomed to failure."

John Dvorak advocated an end to the principal of plagiarism in his 1994 book *Dvorak Predicts*. He wrote: "Papers composed of cut-and-paste excerpts from online services and CD-ROMs are the future of education. None of this should be considered plagiarism, either. In fact, I don't see why the student can't just turn in something written by someone else so long as the other person's name is cited. . . . The student of the future is not the same as an Oxford Scholar of 1895, true. Fact is, the Oxford student would not be able to compete with the computer-augmented student of the future. Times change and we have to change with them."

But technology lawyer Lance Rose said that property should and will be respected and that theft detection would be engaged. "Every time a pirated work is spread to the four corners of the Internet by an anonymous user, software agents will quickly sniff it out," he wrote in a 1994 article for *Wired* titled "The Emperor's Clothes Still Fit Just Fine." "Anonymous infringements will arc across the Net like shooting stars, and disappear from sight just as quickly. Those who want the latest freebie will have to scramble for it before the cops and their software agents go out to sweep up the mess."

Eugene Spafford warned in his chapter in the 1992 book *Computers, Ethics, and Society* that some digital information files must be kept private. "If all information were to be freely available and modifiable, imagine how much damage and chaos would be caused in our real world! Our whole society is based on information whose accuracy must be assured. This includes information held by banks and other financial institutions, credit bureaus, medical agencies and professionals, government agencies such as the IRS, law enforcement agencies, and educational institutions. Clearly, treating all their information as 'free' would be unethical in any world where there might be careless and unethical individuals."

SOME SAY THE CURRENT CONCEPT OF GEOGRAPHY WILL ALSO FALL AWAY

In cyberspace, the current rules of geography may also disappear, according to some seers. First, there's the ability to "be" anywhere in the world at any time through Internet communications. Add to that the potential for further development of virtual-reality worlds, and you can see how these folks get their inspiration.

MIT's Michael Dertouzos, an active participant in the planning for the new National Information Infrastructure (NII), wrote a 1991 *Technology Review* article titled "Building the Information Marketplace" that explained what was to come. He said:

> The NII would make possible a United States where business mail would routinely reach its destination in five seconds instead of five days; where advertis-

ing could be done in reverse, with consumers broadcasting needs to suppliers; where goods could be ordered and paid for electronically; where a retired engineer in Florida could teach high school algebra to a bunch of students in New York City; where a parent could deliver office work to a distant employer while taking care of young children at home; where you might enjoy from your easy chair your choice of a high-definition video movie from the millions produced; where national treasures like the National Gallery could be explored at your own pace and with your interests in mind . . . and on the list goes, limited only by our imagination.

Dertouzos explained that the NII could be seen as a new means of "controlling our personal locality—choosing our working associates, vendors, entertainers, and perhaps even friends, without being limited to those that happen to be physically near." He said that as we see the "importance of physical proximity diminished, every person on the national information infrastructure could assemble his or her own electronic neighborhood."

MIT's William Mitchell wrote in his 1994 book *City of Bits*:

> The Net's despatialization of interaction destroys the geocode's key. There is no such thing as a better address, and you cannot attempt to define yourself by being seen in the right places in the right company. . . . When telecommunication through lickety-split bits on the infobahn supplements or replaces movement of bodies . . . and when telepresence substitutes for face-to-face contact . . . the spatial linkages that we have come to expect are loosened. . . . Buildings will become computer interfaces and computer interfaces will become buildings. . . . We are all cyborgs now. Architects and urban designers of the digital era must begin by reauthorizing the body in space. . . . Once, places were bounded by walls and horizons. Days were defined by sunrises and sunsets. But we video cyborgs see things differently. The Net has become a worldwide, time-zone-spanning optic nerve with electronic eyeballs at its endpoints.

Virtual-reality pioneer Eric Gullichson said in a 1990 interview with the *Washington Times* titled "Cyberspace: Graphics in 3-D Create a Universe," "When this stuff gets implemented and gets widespread, as all the trends indicate it will, what people experience as the ordinary material universe will start to crumble away a little bit. At the very least, virtual reality will become just one of many possible places in which to spend time."

And essayist Nicole Stenger wrote passionately about her vision in a piece written for the 1992 compilation *Cyberspace: First Steps*: "For blind bards as for nearsighted whiz kids, cyberspace will feel like Paradise! Of course don't expect to keep your old identity: one name, one country, one clock. For be it through medical reconstruction or through fantasy, multiplied versions of yourself are going to blossom up everywhere. Ideal, statistical, ironical. A springtime for schizophrenia!"

SPEED AND NEW SOCIAL PRESSURES BRING STRESS

People expressed concerns about the extra stress and social changes that networked communications might create. In his 1991 book *The Saturated Self,* Kenneth Gergen wrote:

> Through the technologies of the century, the number and variety of relationships in which we are engaged, potential frequency of contact, expressed intensity of relationship, and endurance through time all are steadily increasing. As this increase becomes extreme we reach a state of social saturation. . . . Formerly, increases in time and distance between persons typically meant loss. [Today] one may sustain an intimacy over thousands of miles. . . . In effect, as we move through life, the cast of relevant characters is ever expanding. For some this means an ever-increasing sense of stress. . . . At the same time that the past is preserved, continuously poised to insert itself into the present, there is an acceleration of the future. The pace of relationships is hurried, and processes of unfolding that once required months or years may be accomplished in days or weeks. . . . As the future opens, the number of friendships expands as never before.

The heady speed at which a totally networked culture can move was also expressed by Steven Acker of Ohio State University in a 1995 article he wrote for *Computer-Mediated Communication* titled "Space, Collaboration, and the Credible City": "The universal 'we' has lost a sense of rhythm, and is in danger of unbalancing the thought and action cycle that drives creative human behavior. . . . Where traveling through space physically once buffered periods of mental activity, we are squeezing out the inherent rest cycle associated with going to libraries, face-to-face meetings, and going from home to work. . . . The added convenience of telecommunication-based collaboration, the umbrella reason that new technologies are adopted within organizations, carries with it this hidden cost of a loss of pace as it throws us into the vacuum of electronic space."

INFORMATION OVERLOAD VERSUS SNEAKY SOFTWARE AGENTS

Technology developers spent a great many research hours and dollars trying to perfect a way in which "intelligent agents" could be programmed to suit the needs of individual Internet users. Those supporting the idea said these free-roaming software agents would represent human needs in searching the increasingly voluminous Internet and finding and sharing data. People were mostly frightened by this prospect, fearing the lack of security and privacy and associated consequences.

For instance, researcher Sherry Turkle tells in her 1995 book *Life on the Screen* about a student's reaction to a presentation by MIT artificial-intelligence researcher Patti Maes. The student said, "An [intelligent software] agent working for a rival corporation could be programmed to give you very good advice for a time, until your agent came to rely on him. Then, just at a critical moment, after you had delegated a lot of responsibility to your agent who was depending on the double agent, the unfriendly agent would be able to hurt you badly. So the Internet will be the site for intelligence and counterintelligence operations." Another student said. "There's nothing to be afraid of. It's not going to be HAL, but 'Remains of the Day.'" (Referring to the differences between HAL, the nefarious computer in the film *2001: A Space Odyssey*, and the willing-to-please butler played by Sir Anthony Hopkins in the film *Remains of the Day.*)

But even Internet cheerleader Nicholas Negroponte saw potential problems in the use of agents. In a 1995 column for *Wired* magazine titled "Double Agents," he wrote: "Net-dwelling agents are the ones we need to worry about when it comes to privacy. They need to be tamper-proof, and we must find ways to preclude new forms of kidnapping (agent-napping). Sounds silly? Just wait until the courts begin to agonize over whether intelligent agents can testify against us."

GREED COULD ENDANGER FREE SPEECH, FREE WILL

There was a concentrated effort for open access during the early 1990s, and action was taken to assure that access to the Internet would not be dominated and controlled by corporations or media conglomerates. Ken Goffman, who used the pen name R.U. Sirius in his role as editor of the new-age publication *Mondo 2000*, said in a 1992 interview with the Bergen (N.J.) *Record* headlined "Unfolding the Future": "Who's going to control all this technology?

The corporations, of course. And will that mean your brain implant is going to come complete with a corporate logo, and 20 percent of the time you're going to be hearing commercials?"

But some observers saw partial commercialization of the network as a sign of success. In a 1994 newspaper interview with the Wisconsin State Journal titled "Computer Network is Superhighway On-Ramp," National Science Foundation network developer Lawrence Landweber said, "It's supposed to be commercial. The commercialization of the Internet is the proof that what we are doing is worthwhile. It is not just an intellectual exercise for a few professors."

Bill Frezza, a co-founder of DigitaLiberty, wrote on his online site in 1994 that the Internet could never come under the control of any one political or commercial entity.

> Anyone who thinks that this revolution will be dominated by any one power . . . is a fool. This is a bogeyman that has been created by the self-appointed advocates of the information "have-nots" in their efforts to create a new round of complex government dependencies and new industrial policies. The rush to protect us from the 500-channel cable monster is a ruse to justify one last desperate grasp for power. The growth of Cyberspace will not and cannot be planned by any central authorities, government or corporate. It will be a spontaneously self-organizing, continuously evolving, chaotic free-for-all shaped only by the uncoerced choices of the millions of individuals that populate it. And this will make it the perfect crucible for Radical Capitalism.

Science writer Barry Shell is quoted in the 1995 *New Scientist* article "Idiots Guide to the Net": "The Internet, with its open, distributed structure, was designed to withstand a nuclear attack. If it can do that, it can withstand corporate America."

CAN ENCRYPTION GUARANTEE SECURITY?
IS SEAMLESS ENCRYPTION DANGEROUS?

Many observers saw the potential for networked communications to wreak havoc with privacy. In 1992, the *Boston Globe* quoted artist Eric Hughes, creator and performer of "A Prairie Home Computer," after he made the satiric remark, "In the world of the future, people will use low-cost Radio Shack equipment to spy on themselves and find out who they are." And in a 1994 article in *Wired* magazine titled "James Bond R Us" writer Jay Kinney maintained, "Circumstances are conspiring to create a world in which we are all espionage agents, whether we want to be or not. It is not widely recognized yet, but the much-ballyhooed Information Economy is, in fact, an Intelligence Economy where we will all wake up one morning to find ourselves locked into the go-go lifestyle of Spy vs. Spy. Or hadn't you noticed that 'knowledge worker' is just another name for 'intelligence analyst'? In our new Intelligence Economy, we'll all be working for the jokers with the satellites."

The ability to encrypt digital communications became a vital topic, as government and law-enforcement officials feared a future with total privacy could lead to crime, tragedy, and an end to the payment of taxes. Technology writer Steven Levy illuminated the issues involved in a 1993 article in *Wired* magazine titled "Crypto Rebels." "The flood [of cryptography tools] indeed is coming, and the agency charged with safeguarding and mastering encryption technologies is about to be thrust into a cypher age in which messages that once were clear will require tedious cracking—and may not be crackable at all," he wrote.

Privacy advocate and cryptographer Eric Hughes was one of the founders of an active pro-cryptography group called the Cypherpunks. "Cryptography will ineluctably spread over the whole globe, and with it the anonymous transactions systems that it makes possible," he wrote in his 1990s online essay "A Cypherpunk's Manifesto." "For privacy to be widespread it must be part of a social contract. People must come and together deploy these systems for the common good."

When the Clinton administration proposed that computer systems be outfitted with an encryption chip that allowed the government a back door for access, Sun Microsystems cryptographer Whitfield Diffie told a *New York Times* reporter in the 1993 article "Wrestling Over the Keys to the Code": "Electronic communication will be the fabric of tomorrow's society, and we will have daily interaction with intimates that we can only rarely afford to visit in person. By codifying the government's power to spy invisibly on these contacts, we take a giant step toward a world in which privacy belongs only to the wealthy, the powerful, and, perhaps, the criminals."

Georgetown University computer-crime expert Dorothy Denning saw the government's side of the case. She told *Wired* magazine in the 1993 article "Crypto Rebels": "If we fail to enact legislation that will ensure a continued capability for court-ordered electronic surveillance, . . . systems fielded without an adequate provision for court-ordered intercepts would become sanctuaries for criminality wherein Organized Crime leaders, drug dealers, terrorists, and other criminals could conspire and act with impunity. Eventually, we could find ourselves with an increase in major crimes against society, a greatly diminished capacity to fight them, and no timely solution." She added, in a 1993 article for *Communications of the ACM* titled "Digital Communication Must Not Weaken Law Enforcement": "Technology has been drifting in a direction that could shift the balance away from effective law enforcement and intelligence-gathering toward absolute individual privacy and corporate security. Since the consequences of doing so would pose a serious threat to society, I am not content to let this happen without serious consideration and public discussion. . . . The consequence of this choice will affect our personal safety, our right to live in a society where lawlessness is not tolerated, and the ability of law enforcement to prevent serious and often violent criminal activity."

A SENSE OF WONDER AND RESPECT
REIGNED AS PEOPLE MADE DECISIONS

In the early 1990s, a great majority of the American populace was still fairly uninformed about the coming explosion of networked communications. Internet activists like Howard Rheingold asked people to pay attention, step up, and take a stand. "I'm not here with answers; I'm here with questions," he said in a 1993 interview with the *Seattle Times* titled "Author Backs Freedom of Online Expression." "People need to contact their local representatives and get involved. It's up to citizens to demand our part of the deal. If you dismiss this as a bunch of computer nerds, you do your grandchildren a disservice."

Despite all the questions, there was still a sense of wonder and awe about what was transpiring on the network. Technology writer Bruce Sterling wrote in a 1995 magazine article titled "Welcome to Cyberspace": "When I look at the Internet, I see something astounding and delightful. It's as if some grim fallout shelter had burst open and a full-scale Mardi Gras parade had come out. I take such enormous pleasure in this that it's hard to remain properly skeptical."

The following collection of comments from 1992 to 1995 includes some of the most incisive, witty, and revealing remarks made by informed stakeholders and concerned citizens in the first boom years of the Internet, including the voices of Tim Berners-Lee, Lance Rose, George Gilder, William Mitchell, Jay Kinney, Justin Hall, William Gibson, Michael Schrage, Mark Stahlman, Steve Steinberg, Kevin Kelly, and dozens of others. (Keep in mind that authors sometimes wrote under pen names. "Real" names are listed for those whose alternate identities are known; some of these names may be pseudo-nyms.) These statements are divided into "politics," "social," and "economic" sections, although a number of them do cross over to fall under more than one classification.

PERSPECTIVES ON POLITICS

1991 to 1993

Restrictions on the use of information in one country (to protect privacy, for example) tend to lead to the export of that information to other countries, where it can be analyzed and used on a selective basis in the country attempting to restrict it. "Data havens" may emerge reminiscent of the role played by the Swiss in banking, with few restrictions on the storage and manipulation of information. —**Eli Noam**, Columbia University professor, in the 1991 Ithiel de Sola Pool lecture at MIT, as quoted by constitutional scholar Laurence Tribe in the article "The Constitution in Cyberspace," published in *The Humanist* magazine.

The mass media will rapidly lose the vast political power they have exercised for so long. Power will move into constantly shifting communities of shopkeepers, housewives, Yale bulldogs, fruit-juice drinkers, nudists, sandal-wearers, nature-cure quacks, pacifists and phesbian leminists. —**Peter Huber**, a senior fellow of the Manhattan Institute, in a 1992 *Forbes* magazine article titled "Telephone Democracy"

In the Net of tomorrow, the light of criticism must shine everywhere, or secrets which lay hidden will fester into new crises, new weapons, new errors. In an information society, secrecy is the equivalent of cancer. —Technology expert **David Brin**, in a 1992 speech for the Library and Information Technology Association

It is the new standard by which literacy must be judged, and for a time whole nations, but perhaps for government elites, will in effect be made newly illiterate. They need not remain so: This is easier to overcome than hunger or drought. New tasks for future Peace Corps: wire up the global village. —**Robert Coover**, in a 1993 *New York Times* article titled "We are the Wired: Some Views on the Fiberoptic Ties that Bind"

We have the capability of 100-percent privacy. But if we use this I don't think society can survive.—Security specialist **Donn Parker,** in a 1993 *Wired* magazine article titled "Crypto Rebels"

The whole massive, lethal superpower infrastructure comes unfolding out of 21st-century cyberspace like some impossible fluid origami trick. The Reserve guys from the bowling leagues suddenly reveal themselves to be digitally assisted Top Gun veterans from a hundred weekend cyberspace campaigns. And they go to some godforsaken place that doesn't possess Virtual Reality As A Strategic Asset, and they bracket that army in their rangefinder screens, and then they cut it off, and then they kill it. Blood and burning flesh splashes the far side of the glass. But it can't get through the screen. . . . Can governments really exercise national military power—kick ass, kill people—merely by using some big amps and some color monitors and some keyboards, and a bunch of other namby-pamby sci-fi "holodeck" stuff? The answer is yes. Yes, this technology is lethal. Yes, it is a real strategic asset.—Technology writer **Bruce Sterling,** in his 1993 article for *Wired* magazine "War Is Virtual Hell"

This is what will happen: Multiple overlapping of remotely located personalities. . . . Pretty soon, we'll have dozens of alters (intelligent agents) wandering the electronic hallways, meeting other alters, while we wait. . . . The alters ultimately will report back to us about the things they've seen, other alters they've spoken to, the fun they've had. Later, maybe they won't report back to us anymore.—**Curtis Karnow,** in a 1993 article for *Wired* titled "Alters"

As the electronic revolution merges with the biological evolution, we will have—if we don't have it already—artificial intelligence, and artificial life, and will be struggling even more than now with issues such as the legal rights of robots, and whether you should allow your son to marry one, and who has custody of the offspring of such a union.—**Jim Dator,** in a 1993 speech for the WFSF World Conference titled "Dogs Don't Bark at Parked Cars"

For 300 years we have had a scientific ethos that says "information is good"— and the more we know the better. I believe we're heading into an era when there's going to be enormous pressure to block out, to prevent further development of certain kinds of knowledge.—Futurist **Alvin Toffler,** in a 1993 *Wired* magazine interview headlined "Shock Wave (Anti) Warrior"

With the NII looming on the horizon, the question becomes, how ya' gonna keep 'em down on the farm, after they've seen Paree? And cyberspace has lots of provocative "Parees."—Technology writer **Sandy Sandfort,** in his 1993 *Wired* article "The Intelligent Island"

No one group or person is in charge. It's definitely out of control. . . . The role of capital as an editor is being removed.—*Mondo 2000* editor **R.U. Sirius** (real name, Ken Goffman), quoted in a 1993 article in *The Nation* titled "The Whole World Is Talking"

In the globally-linked teleworking virtual judiciary of the future, the judge can be on the beach at Waikiki, the defendant at home in Auckland, his lawyer in Beijing, the prosecuting attorney in Paris, the clerk in Nashville, the probation officer in Pyongyang, the witnesses on the Moon, at L5, exploring life in the superhot plumes of an abyssal trench 40,000 leagues under the sea, climbing Mount Everest, between tennis matches at Wimbledon.—**Jim Dator,** in his 1993 article for *Future Research Quarterly* titled "American State Courts, Five Tsunamis & Four Alternative Futures"

We should have learning centers, neighborhood electronic cottages. . . . It would be easier to get the Pope to become a Buddhist than to get the schools to change.—Education technology expert **Ed Lyell,** in a 1993 *Wired* article titled "Man with a Plan"

Government should play a limited role in building the infrastructure. . . . Private industry, spurred by competition made possible by government policy and motivated by profits (that is, driven by fear and greed), can do a better job than the federal government at delivering communications and information services.—**Rep. Edward Markey** (D-Mass.), in his 1993 article for the book *The Information Revolution* titled "A Legislative Agenda for Telecommunications"

From 1994

It may be unnecessary to constitutionally assure freedom of expression in an environment which, in the words of my fellow EFF co-founder John Gilmore, "treats censorship as a malfunction" and reroutes proscribed ideas around it.—**John Perry Barlow,** in his *Wired* article "The Economy of Ideas"

Any kid with a bright idea should have an opportunity to get on the information highway and sell his or her idea without having to sell out to a monopoly.—**Rep. Ed Markey** (D-Mass.), in a *Wired* magazine article titled "Power PC: As You Read This, Deals Are Being Cut on Radically New Communications Regulations"

The Internet is tougher than you give it credit for. From now on, the struggle will not be over mechanical control of the means of information, but over spin-control of the zeitgeist.—**Bruce Sterling**, quoted in a *London Guardian* article titled "High Anxiety for Hitchhikers on the Infobahn"

Like the printing press, the new computer media will bring forth its own very special ways to think about complexities we have not been able to deal with up to now. . . . Much care has to be taken with design and education in order for the change to be positive. We don't have natural defenses against fat, sugar, salt, alcohol, alkaloids—or media.—**Alan Kay,** in his speech "The Infobahn Is Not the Answer"

Trusting the government with your privacy is like having a Peeping Tom install your window blinds.—**John Perry Barlow,** in his *Wired* article "Jackboots on the Infobahn"

It is precisely because the Net holds the promise of being the most democratizing communications medium in the history of the planet that it is vital we prevent the fearful and the ignorant from attempting to control your access to it.—EFF chief counsel **Mike Godwin,** in a speech at Carnegie-Mellon University after the university banned several newsgroups from campus computers

Monetary transactions will take place through cyberspace if people believe that it is protected. And the same thing is true for medical records. If people believe that there is security and that the network cannot be cracked, then they will use it for financial and other purposes. Superhighway industries will be born only as assurances can be given that there is security on these networks.—**Rep. Edward Markey** (D-Mass.), in a article in *Infosecurity News* titled "Infohighway Security Viewpoints"

[It may already] have become literally impossible for a government to shut people up.—EFF counsel **Mike Godwin,** in an article in *U.S. News & World Report* titled "Pamphleteering in the Electronic Era"

A modem, a PC, and the intent to destabilize might prove a more serious threat to the established order than any military invasion.—Author and social theorist **Douglas Rushkoff,** in his book *Cyberia: Life in the Trenches of Hyperspace*

We've given individuals and small groups equally powerful tools to what the largest, most heavily funded organizations in the world have. And that trend is going to continue. . . . By creating this electronic Web, we have flattened out again the difference between the lone voice and the very large, organized voice. . . . And I don't think it's anywhere near over.—NeXT and Apple CEO **Steve Jobs,** quoted in a *Rolling Stone* article headlined "Looking for the NeXT Revolution: Steve Jobs"

Instead of money going to candidates, money could be given directly to the voters. Instead of tens of millions of dollars in communication vouchers being given to presidential candidates to spend on 30-second television ads, money could be given directly to citizens to spend on information about political candidates. . . . A special class of information agent—the electoral agent—could be given special treatment much like nonprofit organizations. —**James H. Snider,** in his *The Futurist* article "Democracy On-Line"

The left wing is dead. The right wing is dead. Ideology is dead. In place of the stale, 19th-century pre-cyber age ideologies that still provide coinage for "the system" (how could anyone still be proud to be identified as a socialist or a Jeffersonian or a libertarian?), proto-movements are beginning to form to tackle the far more radical politics of cyberspace.—**Mark Stahlman,** in his *Wired* essay "Just Say No to Cybercrats and Digital Control Freaks"

The question is not whether we become a little richer and have a few more toys, but whether civilization, or indeed life on Earth, can survive. What we need is not just any technological innovation, but innovation in the interest of environmental protection of health and safety and of infrastructures that can secure a civilized existence. We must turn our attention to real social problems and away from the twin illusion that technological innovation can achieve economic growth and that growth can solve all our problems.—**Ernest Braun,** a professor at the United Kingdom's Open University, in an article he wrote for *Futures* titled "Can Technical Innovation Lead Us to Utopia?"

Most people think that what you read in *The New York Times* is probably a pretty reasonable representation of world events. But over the Internet, there's a real danger that people would only see what their provider wants them to see or what their religious/ethnic group wants them to see, and you could see a breakdown on the notion of consensus truth.—**Jaron Lanier,** virtual-reality entrepreneur, quoted in an *Electronic Engineering Times* article titled "The 'Nightmare Scenario'"

We have a massive effort under way in the telecommunications conference to say we are going to tell you what to think; we are going to tell you what to do when you go online. Do you know why? I am willing to bet that three-quarters of the Congress do not have the foggiest idea how to get on the Internet. —**Sen. Patrick Leahy** (D-Vermont), in an article on his official Senate website titled "The Role of Department of Justice and Internet Protest"

When cryptography is outlawed, bayl bhgynjf jvyy unir cevinpl!—**John Perry Barlow,** in an online newsletter on the Computer Professionals for Social Responsibility website

In the electronic republic, it will no longer be the press but the public that functions as the nation's powerful "fourth estate," alongside the executive, the legislative, and the judiciary. . . . The emergence of the electronic republic gives rise to the need for new thinking, new procedures, new policies, and even new political institutions to ensure that in the century ahead majoritarian impulses will not come at the expense of the rights of individuals and unpopular minorities.—Former NBC News and PBS executive **Lawrence Grossman,** in his book "The Electronic Republic: Reshaping Democracy in an Information Age"

I'm not against the Internet. I just want people to be more skeptical about it. People are skeptical about nuclear power and genetic engineering and a lot of other areas but they blindly accept the Internet. We techies should be more honest about what computers can do and what they cannot do, or else we are setting ourselves up for a big pie in the face.—**Clifford Stoll,** quoted in a *Financial Times* article titled "Internet Hero Tried to Kick the Habit"

By 2090, the computer will be twice as smart and twice as insightful as any human being. . . . By 2100, the gap will grow to the point at which homo sapiens, relatively speaking, might make a good pet. Then again, the computers of 2088 might not give us a second thought.—**Greg Blonder,** communications scientist, in an essay for *Wired* headlined "Faded Genes"

The future, in my opinion, does not lie in the cable networks. The future is in the Internet, which is 100 percent democratic. . . . Hopefully, the (cable industry) Goliath is going to get hit in the head with a stone, the stone being the Internet.—Auto Channel executive **Bob Gordon,** in a *Chattanooga Free Press* article headlined "Bet on the Well-Connected"

The network is wired to route around trouble.—Networking consultant **Einar Stefferud,** quoted in the *Newsweek* article "The Year of the Internet"

The question of the bit police is a big one, and we know it's big, because we're watching the legal system flop around like a half-dead fish on the dock. They don't know how to handle the new technology.—**Nicholas Negroponte,** in a *Wired* profile headlined "Being Nicholas Negroponte"

If we ever succeed in making machines as smart as humans, then it's only a small leap to imagine that we would soon thereafter make—or cause to be made—machines that are even smarter than any human. And that's it. That's the end of the human era—the closest analogy would be the rise of the human race within the animal kingdom. The reason for calling this a "singularity" is that things are completely unknowable beyond that point. —Sci-fi author **Vernor Vinge,** quoted in the *Wired* article "Singular Visionary: Sci-fi Master/Math Nerd Believes That Machines Are About to Rule the Human Race"

I can imagine proposals that every automobile, including yours and mine, be outfitted with a recorder but also with a transmitter that identifies the car and its location—a future license plate. . . . The black box could record your speed and location, which would allow for the perfect enforcement of speeding laws. I would vote against that.—**Bill Gates,** in his book *The Road Ahead*

It is our belief that the Internet is not controllable, and its standards and protocols should be completely open. If we all sit around and wait for a central body to completely set these standards, then some company—probably Microsoft—will end up creating the de facto standards.—**Jim Clark,** quoted in a *Red Herring* magazine article titled "The Once and Future Kings"

It's so easy to imagine a scenario in which technology is used to get instant (political) judgments from people. If it is used that way, we haven't seen anything yet when it comes to high-tech lynchings.—The Benton Foundation's **Andrew Blau,** in a *Governing* magazine article headlined "Teledemocracy: For Better or Worse"

The Internet is not only going to be a site for the contest of ideas and values, which is what museums do; it could very well be the site of a serious struggle for control of our culture.—**David Ross,** director of New York's Whitney Museum, quoted in a *Wired* article titled "Accommodating the Velocity"

SOCIAL IMPACTS SCRUTINIZED
From 1992 to 1994
Imagine the possibilities when GPS and a cellular telephone can be combined on a cheap chip set and installed almost anywhere. . . . If you want to track migratory birds, prisoners on parole or (what amounts to much the same thing) a teenage daughter in possession of your car keys, you are going to be a customer sooner or later.—**Peter Huber,** senior fellow at the Manhattan Institute in his 1992 article for *Forbes* titled "An Ultimate Zip Code"

I propose to abolish all public grants for schools and colleges and instead give the money directly to families in the form of "micro-vouchers" to be spent on anything that nurtures the spirit and teaches new skills.—Discovery Institute senior researcher **Lewis J. Perelman,** in a 1993 *Christian Science Monitor* article titled "Will Technology Alter Traditional Teaching?"

Computer network and hacker slang is filled with references to "being wired" or "jacking in" to a computer network, "wetware" (the brain), and "meat" (the body). The human body is becoming a hack site, the mythology goes, a nexus where humanity and technology are forging a new and powerful relationship.—Technology writer **Gareth Branwyn,** in a 1993 *Wired* article titled "The Desire to Be *Wired*: Will We Live to See Our Brains *Wired* to Gadgets?"

Experts are talking 2 million characters per second—a forty-fold increase in speed over current technology. If applied to the 55-mph speed limit, it would mean cars zooming at 2,200 miles per hour.—*Cleveland Plain Dealer* reporter **Keith Epstein,** in his 1993 article "300-Lane Information Superhighway Brings Data, Entertainment to Your Door"

In the future, computers will mutate beyond recognition. Computers won't be intimidating, wire-festooned, high-rise bit-factories swallowing your entire desk. They will tuck under your arm, into your valise, into your kid's backpack. After that, they'll fit onto your face, plug into your ear. And after that—they'll simply melt. They'll become fabric. . . . Fabric and air and electrons and light. Magic handkerchiefs with instant global access. You'll wear them around your neck. You'll make tents from them if you want. They will be everywhere, throwaway. Like denim. Like paper. Like a child's kite. This is coming a lot faster than anyone realizes.—**William Gibson,** in a 1993 speech for the National Academy of Sciences Convocation on Technology and Education

Video conferencing bears a terrifying promise: Distance will no longer be an excuse for not attending meetings.—**Steve Steinberg,** in a 1994 *Wired* article titled "Hype List: Video Conferencing"

The Internet will be to women in the '90s what the vibrator was to women in the '70s. It's going to have that power.—Cyberporn magazine editor **Lisa Palac,** quoted in a 1994 *GQ* magazine article headlined "Pigs in (Cyber) Space: On the Internet, as in Life, Women Put Up with the Damnedest Things"

Telemolesters will lurk. Telethugs will reach out and punch someone.—**William Mitchell,** in his 1994 book *City of Bits*

You really do not want the Library of Congress on your desk. What you want is to be able to get at the Library of Congress from your desk. The reason you don't want it on your desk is because as it gets out of date you have to worry about maintaining the Library of Congress!—**Danny Hillis,** inventor of massively parallel computing, quoted in a 1994 *Wired* magazine article titled "Kay + Hillis: *Wired* Brings Together Two Legendary Minds"

Networks at . . . different levels will all have to link up somehow; the body net will be connected to the building net, the building net to the community net, and the community net to the global net. From gesture sensors worn on our bodies to the worldwide infrastructure of communications satellites and long-distance fiber, the elements of the bitsphere will finally come together to form one densely interwoven system within which the knee bone is connected to the I-bahn.—**William Mitchell,** in his 1994 book *City of Bits*

From 1995
Every family will have its own mailing list carrying contributions from its members. At that point—actually long before it—we will have to triage our mail still further. While I have had to do very little of this so far, I sense that the rules will be something like this: friends over strangers; family over friends; and within those categories, the geographically or chronologically close over the distant.—Technology writer **Fred Hapgood,** in a *Wired* article titled "Persistence of Locality"

Paradoxically, the superhighway may connect us more to other people of similar interests and beliefs. But we'll have less communication with those who are different. Socially we may find ourselves returning to a form of tribalism, as we separate ourselves along group lines—racial, ethnic, ideological—choosing access to only the information that speaks to our identities and beliefs. There may be more avenues for individual speech but fewer for robust, wide-open debate.—Author **Les Brown,** quoted in a *Seattle Post-Intelligencer* article headlined "Scholars Try to Measure the Impact"

What some people call hate crimes are going to increase, and the networks are going to feed them. I believe in the First Amendment. But sometimes it can be a noose society hangs itself with.—**Carlton Fitzpatrick,** branch chief of the *Fi-*

nancial Fraud Institute, quoted in a Wired magazine article titled "Policing Cyberspace"

There are criminals in the world, and some of them are programmers. With computer networks, they have an amplifying effect that they've never had before. If I were a criminal with a gun, I might attack one person. But with a computer network, I can attack a million people at a time. It's like an atomic bomb.—**Eric Schmidt,** chief technology officer for Sun Microsystems, in a *New York Times* article headlined "Computers Beware! New Type of Virus Is Loose on the Net"

Network mail, even decade-old e-mail, lacks warmth. The paper doesn't age, the signatures don't fade. Perhaps a future generation will save their romances on floppy disks and Internet Uniform Record Locators. Give me a shoebox of old letters.—Internet critic **Clifford Stoll,** in his book *Silicone Snake Oil*

I had (and still have) a dream that the Web could be less of a television channel and more of an interactive sea of shared knowledge. I imagine it immersing us as a warm, friendly environment made of the things we and our friends have seen, heard, believe or have figured out.—**Tim Berners-Lee,** in a speech in honor of the 50th anniversary of Vannevar Bush's visionary article "As We May Think" at MIT

I have had a vision of the future. In it, we live in a virtual-reality environment and we spend all our waking hours as employees of the U.S. Electronic Postal Service. Wake me when it's over.—**Sorell Reisman,** in a column for *IEEE Software Journal* titled "I Have Seen the Future; It is Flooded with E-mail"

I dream that we would see involvement of students in a variety of virtual communities, engagement of teachers as lifelong learners and researchers, use of online student portfolios and assessment, and regular use of the full range of network resources. . . . A whole new class of information organizers and intelligent scaffolding software will help students structure their ideas and reflect on their thinking. These tools will finally help realize the promise of metacognition, understanding how you think and learn. . . . The result could well force a total rethinking of what can be taught and when.—Education technology expert **Bob Tinker,** president of Concord Consortium, in a white paper titled

"The Whole World in Their Hands," commissioned by the U.S. Department of Education's Office of Educational Technology

I'm a future hacker; I'm trying to get root access to the future. I want to raid its system of thought. Grrr. Machines disappoint me. I just can't love any of these wares, hard or soft. I'm nostalgic for the future. We need ultrahigh res! Give us bandwidth or kill us! Let's see the ultraviolet polka-dot flowers that hummingbirds see, and smell 'em like the bees do. And crank up the sensorium all across the board. . . . I think tech will solve all our problems, personal and scientific. Girls need modems.—**St. Jude** (real name, **Judy Milhon**), quoted in a *Wired* article titled "Modem Grrrl: Future Hacker St. Jude Has Some Advice for Women Who See Technology as a Problem"

I think the best paradigm for the Internet is to forget that it's a technology and pretend that it's a language. . . . [The] Internet behaves more like the English language than like a technology. . . . I can scarcely believe what I'm seeing nowadays. There's something primally exhilarating about being in a bus without brakes.—**Bruce Sterling,** in an e-mail interview with Telecommunications International titled "Dropping Anchor in Cyberspace"

No matter what circumstances we face or predilections we harbor, the business of living is love. Getting love and keeping love. Manufacturing love. Making love. Making love stay. And no worldwide web of cool chips and hot wires is going to change that. So just shut up about your Brave New World, bub; we've all still got to live in the frightened old one.—**Philip Mart,** in an article in the *Arkansas Democrat-Gazette* headlined "Cyber Utopia a Mirage"

I'm looking forward to the day when my daughter finds a rolled-up 1,000-pixel-by-1,000-pixel color screen in her cereal packet, with a magnetic back so it sticks to the fridge.—World Wide Web inventor **Tim Berners-Lee,** in an *Information Week* article titled "The Internet—Where Is it All Going?"

Jewelry that is blind, deaf, and dumb just isn't earning its keep. Let's give cuff links a job that justifies their name. . . . And a shoe bottom makes much more sense than a laptop—to boot up, you put on your boots. When you come home, before you take off your coat, your shoes can talk to the carpet in preparation for delivery of the day's personalized news to your glasses. —**Nicholas Negroponte,** in a *Wired* column titled "Wearable Computing"

We just need to figure out why telephones should merge with computers in the first place. True, the combination could help telemarketers. But for most of us, it's as useful as a computerized toaster.—**Steve Steinberg,** in a *Wired* article headlined "Hype List: Computer Telephony"

In the end, reputation, reliability, reality—those will dominate the virtual world, as they rule the real.—*Wired* magazine publisher **Louis Rossetto,** in an e-mail he sent to *Wired* staff members, as reported by Gary Wolf in his book *Wired: A Romance*

The bottom line when it comes to kids, sex and the Internet is that no matter what laws we pass and what high-tech solutions we devise, the three of them together will never be less volatile than the first two alone. We can mitigate but not eliminate the drawbacks of high tech; there's no way to get its benefits without them.—Technology writers **Katie Hafner and Steven Levy,** in a *Newsweek* article titled "No Place for Kids? A Parents' Guide to Sex on the Net"

You'll watch "Gone With the Wind" with your own face and voice replacing that of Vivien Leigh or Clark Gable. Or see yourself walking down a runway at a fashion show wearing the latest Paris creations adjusted to fit your body or the one you wish you had.—**Bill Gates,** in his book *The Road Ahead*

When television is digital, it will have many new bits—the ones that tell you about the others. . . . Today's TV set lets you control brightness, volume, and channel. Tomorrow's will allow you to vary sex, violence, and political leaning.—**Nicholas Negroponte,** in his book *Being Digital*

The heavily promoted information infrastructure addresses few social needs or business concerns. At the same time, it directly threatens precious parts of our society, including schools, libraries and social institutions. No birds sing.—**Clifford Stoll,** quoted in a interview with *The Buffalo News* titled "Time to Exit the Information Superhighway?"

What we're seeing now is something called the armored cocoon, where people are just too scared to go out into their neighborhoods or to the mall, so they invest in expensive security and the outside world comes to them. All services are based around home delivery, the Internet plays a huge part in the armored cocoon and with the rise of cybertechnology, fewer people are going to want to leave their homes when they can have more fun in cyberspace—I'm sure that some people will live "out there" all the time.—Futurist **Faith Popcorn**, in an article in *The London Independent* headlined "Log In, Stay In, Cash Out: Are You Ready for the 21st Century?"

The Internet is more like a social space than a thing so that its effects are more like those of Germany than those of hammers.—Media theorist **Mark Poster**, in a scholarly paper titled "Cyberdemocracy: Internet and the Public Sphere"

I want to know what we are becoming if the first objects we look upon each day are simulations into which we deploy our virtual selves.—**Sherry Turkle**, in the introduction pages of her book *Life on the Screen*

ANTICIPATED ECONOMIC IMPACT

From 1992 to 1994

The value of fixed information will decline greatly. Those individuals and organizations that are responsible for creating the information will have more direct control over the distribution and obtain a greater share of the value. . . . In the near future there will be a new cottage industry in the creation, organization, and distribution of information. Electronic marketplaces open to the public will be one of the biggest societal applications of this technology at some point in the future. . . . New products could be brought to market for far less than is now possible, and without the large marketing budgets that only very large organizations can afford. Electronic Marketplaces offer the opportunity to lower the entry barriers to small-scale entrepreneurship, and to reverse the trend of greater concentration of wealth in fewer and larger organizations.—**Starr Roxanne Hiltz and Murray Turoff**, in a paper they presented at a 1992 workshop titled "Rights and Responsibilities of Participants in Networked Communities"

A new industry is being created in America. We are beginning to create massive Data Libraries. . . . No one place will contain the information. Rather all information sources are being made available in electronic form through the Internet.—Education administrator and technology consultant **Ed Lyell**, in a speech he delivered at a summer training program for the New York State Deans of Education Conference titled "Merging High Tech with High Touch"

The value of information about information can be greater than the value of the information itself.—**Nicholas Negroponte,** in a 1994 *Wired* magazine column titled "Less Is More: Interface Agents as Digital Butlers"

The real future of advertising and marketing may rest in the new ecology and the interrelationship between genes and memes. . . . What are the memes that urge us to purchase and consume? . . . Forget demographics and psychographics. Think memegraphics. . . . If an Emilypost virus were to monitor civility, why not a Coca-Cola virus circulating through the Net and attaching itself to any program using the phrase "The Real Thing?" . . . Of course, it would be nothing short of an art to design ad viruses that were just appealing enough to be noticed and remembered, but never so intrusive as to be obnoxious and alienating . . . some netcrawlers will be virulently anti-advirus. They'll want Lysol-like software to scour and disinfect adviruses from any program before it can be displayed. We'll see an epidemiological battle between the forces of digital commercialism and purists who think commerce has no place on The Net. Of course, we all know who'll win that battle, don't we?—MIT Media Lab fellow and *Adweek* columnist **Michael Schrage,** in an article he wrote for *Wired* headlined "Adviruses, Digimercials, and Memegraphics: The Future of Advertising Is the Future of Media"

From 1995

Congress worries about the information-rich versus the information-poor, but most of its members probably don't realize that computers can cost less than bicycles.—**Nicholas Negroponte,** in a column for *Wired* headlined "Affordable Computing"

The Net is for communications between people. Don't shove products in our faces. We'll know where to find you if you have what we want. If you make good stuff, we'll hear about it. If you want to spread the word, give folks free shit without obligation and ask them to tell their friends. If you want to advertise, do so in context—context not of demographics, but of content.—**Justin Hall,** in an essay from his website titled "I Am Not My Habits: On Our Guard Against Targeted Advertising"

I feel pretty cranky . . . kick-the-neighbor's-cat cranky. . . . I'm weary of the pretension and condescension of those who believe nirvana is here, frying quietly in the circuits of some magic box. I'm sick of the cyber hype. . . . It's more Edgar Cayce than Carl Jung out there, Boopsie. It will not last in this form, no matter how many obsessives are out there. . . . Besides, the telecommunications companies have caught on and sooner or later they're going to impose some capitalist rationality on the whole system. It's going to cost more—a lot more.—**Philip Mart,** in an article from the *Arkansas Democrat-Gazette* headlined "Cyber Utopia a Mirage"

The record store of the future will have a bed to lie in, a well-stocked refrigerator, a bathroom with a shower and maybe even a couch or two. The record store of the future will be your home. . . . Computer users will simply click their mice on albums they're interested in, then sit back while the music is transferred, or downloaded, through high-tech telephone lines to their CD drives, which will capture the music on compact disks.—**Neil Strauss,** in a *New York Times* article headlined "Records of the Future: At Your Fingertips"

[Encryption] just has to cost more to break than it would cost the bad guys to bribe your cleaning lady.—Technology lawyer **Stewart Baker,** quoted in an article in *The (London) Independent* headlined "Fears Grow of a Crime Spree in Cyberspace"

Maybe one day, you'll sit at home and browse through the biggest jukebox in the universe. . . . We will have entered a new era in which music flows into our homes like water.—Unix inventor **Ken Thompson,** quoted in a *Wired* article titled "The Father of Unix Has Developed a New Technology that Could Mean Never Having to Buy a CD Again"

Are we headed toward a world filled with anemic drones, laboring away at sterile keyboards, never taking a moment to sniff the ragweed, never twisting an ankle while tossing a Frisbee to their flea-ridden dogs? Well, we might be. America, at least, has been headed there for some time, roughly since the invention of the fluorescent tube. The Internet, though, is just a symptom of our technological cocoonery, not the root cause.—Media commentator **Dinty Moore,** in the 1995 book *The Emperor's Virtual Clothes: The Naked Truth About Internet Culture*

Most things that succeed don't require retraining 250 million people.—**Waring Partridge,** AT&T's vice president for multimedia strategy, referring to the need to keep networked products simple, in a *Wired* article titled "What Does a Nobel Prize for Radio Astronomy Have to Do with Your Telephone?"

Investors poured a lot of money into the Internet during 1995, but very little leaked out. Everyone will realize this suddenly in January when financial results are tallied. A hurried search for greater fools to absorb projected continuing losses won't pan out this time.—Ethernet inventor **Robert Metcalfe,** in an *InfoWorld* column headlined "From the Ether: Predicting the Internet's Catastrophic Collapse and Ghost Sites Galore in 1996"

We will see a shift toward information services instead of information hoarding. For instance, it would not be surprising if much of what is sold today as "products"—recorded songs, books, films—become no more than cheap promotional tools for premium services, such as live online concerts and direct interactions between audiences and artists. . . . There will simply be more money in helping people use information than in metering the stuff out. —Technology lawyer **Lance Rose,** in a *Wired* article titled "The Emperor's Clothes Still Fit Just Fine"

I was thinking about ecological computing. . . . Pretty soon we're going to have to grow software, and we should start learning how to do that. We should have software that won't break when something is wrong with it. As a friend of mine once said, if you try to make a Boeing 747 six inches longer, you have a problem; but a baby gets six inches longer 10 or more times during its life, and you never have to take it down for maintenance.—**Alan Kay,** in the 1995 *Wired* article "Kay + Hillis"

Chief among the new rules is that "content is free." . . . The creator who writes off the costs of developing content immediately—as if it were valueless—is always going to win over the creator who can't figure out how to cover those costs. The way to become a leading content provider may be to start by giving your content away. This "generosity" isn't a moral decision: It's a business strategy.—Technology consultant **Esther Dyson,** president of Edventure Holdings, in her *Wired* magazine article titled "Intellectual Value: A Radical New Way of Looking at Compensation for Owners and Creators in the Net-Based Economy"

Information is as basic to American households as the proverbial basket of groceries. The designers and marketers of new computer, online and interactive services can succeed by solving the information problems that time-sensitive consumers face.—**Thomas Miller,** vice president of the emerging technologies group of Find/SVP, in his *American Demographics* article "New Markets for Information"

When you turn on your computer, your wallet will boot up on your desktop. From Internet tolls to pay-as-you-play software, you will be assessed endless bursts of miniscule fees by a dizzying array of personal purveyors.—**Justin Hall,** in a 1995 essay on his website titled "Digital Cash: Deft Donations or Surf-Stopping Tithing?"

The "Highway" Metaphor

Finding a way to tell (and sell) how the Internet could be changing lives

How do you sell an expensive new idea to people—convince them to buy in? You tell them about all the good things that will ensue if they support it. How do you sell the general public on adopting an extremely complicated and expensive new communications system? You simplify it by finding a fitting metaphor—a familiar, comparable image to which people can relate.

Before they were perfected and popularized, the telegraph, telephone, radio, and television had been dreamed of and discussed. So, too, the Internet—a network allowing the world to communicate large amounts of information at great speed—had been envisioned many times throughout the latter half of the 20th century.

Research scientist Vannevar Bush, dean of the Massachusetts Institute of Technology and a top presidential adviser during World War II, was the first to seriously outline to the general public a detailed idea of what we would later come to recognize as networked computers using hypertext. It was all described in the 1945 *Atlantic Monthly* article "As We May Think." In it, Bush proposed the development of a machine to help researchers make sense of the vast amount of information being published.

FIRST THERE WAS THE MEMEX, THEN INFORMATION THEORY

Bush called his machine the "memex," and explained that it would help people find information, writing of "a new profession of trailblazers, those who find delight in the task of establishing useful trails through the enormous mass of common record."

The work of Bush inspired the future developers of hypertext, including Ted Nelson, Douglas Engelbart, and Andreis Van Dam. Bush basically described the type of linking now common in weblogs (online outpourings of personal information that include links to pertinent related information) and wikis (which allow a person to publish online information that others can come along and modify, add to, and change online).

The phrase *Information Theory* had its origins at the Bell Labs in the 1940s. Claude Shannon, a student of Bush at MIT, brought his work on the efficient transmission of communications to Bell in 1941. His paper, "The Mathematical Theory of Communication," began the field of Information Theory. Because of Shannon's research, Boolean arithmetic became the basis for the design and operation of computers. The term *Information Age* is said to have originated as a sales tool in the 1960s, when AT&T's publicity department made liberal use of the phrase.

In 1961, John McCarthy, the computer-science professor at Stanford University who coined the term *artificial intelligence,* spoke of the ideal "public-utility information system" made up of computers connected by public telephone networks. He said it would interact with consoles in homes, schoolrooms, and offices. During that decade of the 1960s, the first theories behind the working Internet began to emerge in the work of J.C.R. Licklider, Paul Baran, Leonard Kleinrock, Donald Davies, and Lawrence Roberts.

SMITH FIRST WROTE ABOUT A COMMUNICATIONS HIGHWAY

The first known tie-in between the idea of a highway system and a communications network was made by Ralph Smith in a 1971 article in *The Nation* in which Smith also coined the phrase *The Wired Nation.* At the time, Smith was writing about the need for the Federal Communications Commission to alter regulatory barriers that were hindering the development of cable television. That same year, ARPANET connected the beginning of the Internet in the United States, with just 15 interconnected nodes, most of them at research centers. Over the following decade, the computer network grew.

By the late 1980s, Al Gore Jr. (a member of the U.S. Senate at the time and the son of Tennessee Senator Al Gore Sr., a co-author of the act that established the interstate highway system) was associating the idea of the new, networked communications medium with the U.S. network of highways. In 1986, Gore spoke out in favor of the National Science Foundation Authorization Act, saying, "We will need a telecommunications highway connecting users coast to coast, state to state, city to city." In 1989, Gore made a speech in which he said, "High-capacity fiber-optic networks will be the information superhighways of tomorrow."

The word *infrastructure* has long been used for common systems shared and maintained by a culture. In a 1991 article for MIT's *Technology Review,* Michael Dertouzos of MIT said, "Computers will become a truly useful part of our society only when they are linked by an infrastructure like the highway system and the electric power grid, creating a new kind of free market for information services." In 1993, the United States passed the National Information Infrastructure (NII) Act and printed the government report "National Information Infrastructure: Agenda for Action." The official NII phrase was rather stiff, so for public-relations applications the highway metaphor took hold and grew. The Clinton administration continued popularizing the catchy phrase *information superhighway* through the speeches and writings of then–Vice President Gore, and the Internet boom and accompanying media attention sent it through the roof.

TRANSPORTATION NETWORKS WERE ONCE THE MAJOR COMMUNICATION NETWORKS

Of course, the history of communications is directly tied to that of transportation. The first long-distance communications were oral and, later, printed reports, carried on foot and by early wheeled transport such as wagons and by boat. The Roman Empire of the 5th and 6th centuries proved the importance of good highways in holding together a republic made up of far-flung territories. Leaders in times since then have seen transportation and communications networks to be key elements of success in any democratic system. (Tyrants, on the other hand, see the value in keeping people isolated.) In the earliest years of the 19th century, President Thomas Jefferson made it a goal of his administration to develop a national transportation/communications system of canals and roads. A few decades later, railroad networks were added to the national equation.

From 1816 to 1840, more than 3,300 miles of artificial waterways were dug in the United States at a cost of more than $120 million. The earliest version of U.S. Highway 40, the Cumberland Road from western Maryland through Columbus, Ohio, to the Mississippi in Missouri—also known as the National Road—was started in 1815 and completed in the 1850s.

The first railroad in the United States—the Charleston and Hamburg—began regular service in 1831, and 10 years later more than 3,300 miles of track had been laid, mostly in the east. By 1860, hundreds of railroad companies were operating on 30,600 miles of track, and on May 10, 1869, a golden spike was driven into the ground in Utah as the world's first transcontinental railroad was completed. Of course, the U.S. transportation networks grew to include aviation in the early 20th century.

TRANSPORTATION NETWORK DEVELOPMENT IS REFLECTED IN THE INTERNET

Between 1800 and 1870, the United States spent hundreds of millions of dollars to develop the most advanced transportation networks in the world thanks to social, political, and economic partnerships forged among federal, state, and local governments and private business (just as the development of the Internet was a collaborative work completed by the same groups in the 20th century).

Turnpikes, toll roads, railroads, and waterways were constructed, and the building process proved to be one of the best training grounds ever for engineers (just as the development of the Internet became a great training ground for electrical engineers and computer scientists in the 20th century).

When Congress passed its most massive road-building measure in 1956, it was called the Interstate and National Defense Highway Act. The cost of this interstate highway system has been estimated to be at least $38 billion. The point was to construct 41,000 miles of toll-free highways, and they were financed—at least in part—because the government during the Eisenhower administration was fearful of a military conflict in the years of the Cold War (just as the development of the Internet was first financed during the Eisenhower administration in preparation for conflict due to the Cold War).

Over time, transportation networks have gradually altered the ways people relate to one another, the methods by which they conduct business, and the laws by which they govern and are governed; our continually improving waterways, highways, railroads, and air routes have been responsible for changes in property law, the reduction of consumer prices, the easing of migration from place to place and job to job, increased variety in and quality of the available supply of consumer goods, and a tighter network of interpersonal connectedness (just as the development of the Internet is changing economic, social, and political systems).

ACCELERATING ONTO THE INFORMATION SUPERHIGHWAY

Putting things in historical perspective, it's easy to see the computer as the equivalent of the automobile and to see the Internet as the highway system.

As with the computer, in the early days of the automobile the first cars were too complicated for nonmechanics to operate, they were relatively expensive, and they were unreliable, breaking down often. As with the Internet, there weren't many paved roads to travel when the first cars were available, and travel speeds were slow. Both the vehicle and the network had to be improved in order for this system to be embraced by the general public. They were, and the automobile and open highway became a vital part of the American structure and psyche.

When government and industry leaders of the late 20th century were trying to encourage the development of the newest democratic/commercial network, the new popularity of desktop computing and the networking expansion and bandwidth improvements of the 1980s and 1990s could be seen as analogous to the development of reliable, affordable cars and the completion of the interstate highway system.

PLAYING POLITICS WITH THE POTENTIAL
OF A NEW COMMUNICATIONS NETWORK

The "highway" references really picked up during the presidential campaign year of 1992, when George Bush faced off against Bill Clinton. The Clinton team had selected Gore, the most technology-oriented man in the U.S. Senate, as its candidate for vice president. Rhetoric about the information infrastructure began to fly, but it was put into highway terminology.

According to a 1992 *Washington Post* article headlined "The Tekkie on the Ticket," Clinton introduced his technology policy to the press, saying, "I will give our vice president, Al Gore, the responsibility and authority to coordinate the administration's vision for technology and lead all government agencies, including research groups, in aligning with that vision. . . . Such a network could do for the productivity of individuals at their places of work and learning what the interstate highway system of the 1950s did for the nation's travel and distribution system." Gore was also quoted in the *Post* article saying: "Just as every home in America is now linked to the rest of the country by a driveway that goes to a street, that goes to a highway, that goes to an interstate, we want every home and business in America to be linked by an information highway into a national network that will ultimately be . . . invisible to the user in the sense that the information will be there at the desktop terminal."

After Bush lost the election, technology consultant George Gilder delivered a defense of conservatives' contributions to the establishment of the information highway in a 1993 *Wired* magazine interview titled "When Bandwidth is Free" in which he said: "Bush did virtually everything that Clinton promises to do, and because Bush has done it already it doesn't leave Clinton much room except to play cock-a-doodle-do. He'll get up on the post and crow as the marvelous sunrise technologies come blindingly to the fore during his administration. They're going to have 50,000 technology programs and lo and behold, a million technologies will bloom and they will take credit for it all."

In March of 1993, a writer for *Congressional Quarterly Weekly Report* interviewed Gore about his responsibility as Clinton's crafter of Internet policy. Gore said:

> The existence of a premier information marketplace in the United States will give our companies and our citizens an advantage in worldwide competition. There will be many other developments that are currently invisible to us. Just as the development of the Interstate Highway System led to the creation of McDonald's hamburgers, Holiday Inn, and a thousand other new commercial developments that would have been impossible without the Interstate Highway System, in the same way we will see the emergence of information services on a nationwide basis that will be extremely profitable and nearly ubiquitous. . . . The number of miles dwarfs that of the interstate. But they would not exist except for the clear direction, the definition of standards, the clear idea of what an interstate highway is, what it looks like, where it is, what it means. And then everybody else can sort of aim toward that reality. And that's what's happening [with the Internet] and we anticipated that.

Microsoft mogul Bill Gates, the developer of the operating systems used on most of the personal computers in the world, was late arriving to the Internet-hype party, but he hopped onboard the highway bandwagon as well, telling a *Boston Globe* reporter in the 1993 article "Bill Gates, Evangelist," "There is this nonexistent business called 'information highway.' It's got zero dollars in revenue and a lot of uncertainty of when it will emerge. I happen to be someone who believes very much in this business. . . . The mania is in full force, and if you want to be way out in front on this thing you have to move full-speed."

TRAFFIC JAMS, FATALITIES, POLLUTION, AND OTHER PROBLEMS

In the early 1990s, the Internet–Interstate highway analogy was criticized by some for being overused, often in an exaggerated manner, to hype a new communications form. The hyping of a new communications technology is nothing new; people made many of the same claims about the telegraph, radio, television, and cable television that they began making about the Internet in the 1990s. In the 1930s and 1940s, for instance, commercial television was sold to the public as a tool that could reinvigorate family ties and make cities unnecessary.

While the information highway analogy did work well, and there is no doubt it helped sell the concept of networked communications to the masses, it was also a useful tool for those who saw the down side of the technology. Those who opposed the build-up of the Internet took great pleasure in pointing out the detrimental aspects of blanketing the world with highways.

In 1993, Bill McKibben, an author whose book *The Age of Missing Information* involved his watching an entire day of television broadcast on more than 100 channels, was quoted in a *New York Times* article:

> I've heard happy salutes to the coming "information superhighway." The phrase—beloved of the president, too—seems instructive to me. For what is a superhighway? It is fast, and this technology will surely be fast. But it is also flat, uninteresting, repetitive. A path provides you with immense amounts of information about a place—walking it, you feel the topography, the weather. You have time to look at what's around you, to talk with anyone you meet. Even a back road—a blue highway—offers a lot of information, leaves you with some sense of the country you've driven through. But a highway makes sure you can't even stop at a restaurant that might serve something local. So with the communications superhighway, I think.

Computing pioneer Alan Kay expressed his concerns about the new medium in a speech at UCLA in 1994 that was reprinted in *Wired* magazine under the headline "The Infobahn Is Not the Answer." In his speech, Kay said, "Another way to think of roadkill on the information highway will be the billions who will forget there are offramps to destinations other than Hollywood, Las Vegas, the local bingo parlor, or shiny beads from a shopping network. Not couch potatoes but mouse potatoes! It's not the wonderful things they could do with new media, it's what they will be convinced they should do. This is a new tragedy in the making. No democracy that is less than 10 percent literate can survive in the driving forces of society."

Social commentators Richard Sclove and Jeffrey Scheuer looked for all and thoroughly illuminated most if not all of the negatives in their 1994 *Washington Post* essay titled "The Ghost in the Modem: For Architects of the Info-Highway, Some Lessons from the Concrete Interstate." They wrote: "It is easy to romanticize new technology. The popular arts glorified life on the highway. People read Jack Kerouac's 'On the Road,' watched 'Route 66' on television and recall the Merry Pranksters' psychedelic bus-capades during the '60s. In fusing alienation and rebellion with youthful exuberance, each of these foreshadows contemporary cyberpunk culture. Yet real-life experience on the interstate is mostly banal and uneventful. McDonald's, Pizza Hut, and Wal-Mart look about the same wherever you exit."

Sclove and Scheuer see danger ahead on the tech highway, adding: "Rush-hour traffic jams, gridlock, garish plastic-and-neon strips, high fatality rates, air pollution, global warming, depletion of world oil reserves—have we forgotten all of the interstate highway system's most familiar consequences? Comparing the electronic and asphalt highways is useful—but mostly as a cautionary tale. . . . sweeping geographic relocations and accompanying social transformations seem probable. And the risk of inequity in contriving and distributing electronic services—or conversely imposing them where they are not wanted—is clear."

By 1994 the highway phrase had become so overused that humor columnist Art Buchwald wrote, "I am recommending a five-day jail term for anyone who uses the term." In 1994, a *PC World* columnist asked readers to send in alternate names for the "information highway." The column with the responses, "Abort, Retry, Fail: Renaming the Info Highway" included Kevin Kwaku's suggestion that while it's a bad metaphor, it could be a good acronym, standing for Interactive Network For Organizing, Retrieving, Manipulating, Accessing (and) Transferring Information On National Systems, Unleashing Practically Every Rebellious Human Intelligence, Gratifying Hackers, Wiseacres, And Yahoos.

STRUGGLING WITH GRIDLOCK ON THE INFOBAHN:
SHOULD IT BE A TOLL ROAD?

In the early 1990s, bandwidth was limited, and gridlock was a commonly used term because the Internet could work slowly or seem dead-still as more and more people accessed a network that was not yet ready to handle the demands of a new outpouring of travelers. Touring, working, or playing online could be an incredibly frustrating process. Complaints about this problem were often couched in "highway" analogies.

In the essay "Meeting the Challenges of Business and End-User Communities" in the 1995 book *Public Access to the Internet,* Internet analyst and author Daniel Dern wrote,

> [The new-user] group represents our biggest challenge: to find ways to let them educate themselves in the parking lots of cyberspace, rather than on the main roads where other users are trying to get work done. . . . It only takes a percentage of a percentage of new, well-meaning but ill-directed users (not to mention the less-than-well-meaning) to disrupt the activities of thousands, even millions, of regular Internet users. . . . With public-access Internet sites, anyone with a personal computer and a modem can become an Internet user. This is the equivalent of being able to buy an automobile and go driving without having to take a driver's education course, pass a test or become licensed. . . . It creates the reality of tens of thousands of users set loose on the "Internet on-ramp" and raring to go. These users don't necessarily do any harm, but they can place enormous, unanticipated loads on Internet services.

The growing online hordes had Internet Architecture Board research scientists scrambling to expand bandwidth, which they did accomplish by the late 1990s. In the meantime, other people were trying to figure out ways to organize the traffic and simultaneously make it cost-efficient.

University of Michigan economics professors Jeffrey K. MacKie-Mason and Hal R. Varian came up with one of the most-discussed plans, outlined in the 1995 book *Public Access to the Internet.* Noting that open highways are the analogy, but that the world has many toll roads, they wrote:

> No one argues that use of all transportation networks should be free. The interstate highway system might be viewed as the one-size-fits-all, universal-access option (for those who can afford cars), with the option to pay for using

a mode with a different combination of service characteristics. Likewise, a government might want to provide universal, free access to a baseline set of Internet transport services, and allow charges for usage of other services above a threshold. Appropriate free services might include plain-text e-mail (with lower priority when the network is congested) but not guaranteed, zero-delay multimedia broadcast. Universal access and a base endowment of usage for all citizens could be provided through vouchers or other redistribution schemes.

Fortunately, within a few years of these comments, an increase in bandwidth and other system efficiencies made it possible for traffic of a greatly increased magnitude over that of the early 1990s, enabling the use of the Internet to basically remain "free" of charge.

DISCUSSING THE POTENTIAL FOR A "DIGITAL DIVIDE"

Some people predicted that the Internet would cause a "digital divide," causing a bigger separation between the rich and the poor in regard to skill levels and potential opportunities. Arguments on both sides of the issue made references to the information highway.

Communications researcher Oscar Gandy wrote a 1994 article for the journal *National Forum* titled "The Information Superhighway as the Yellow Brick Road." In it, he predicted that the information revolution wrought by digital networks would be detrimental. "The Yellow Brick Road cannot take me anywhere I would like to go; the information superhighway will do no better," he wrote. "Competitive pressure and economic rationalism require firms to focus attention on individuals, or groups, or even communities that their best intelligence suggests have the greatest potential for profit. This means, however, that certain individuals, because their profiles suggest a lower profit potential, will be ignored or bypassed as they stand by the highway trying to hitch a ride to the good life."

That same year, Gene Wolf took the opposite view, painting the Internet as a liberating technology in a speech for the American Library Association's annual conference. "[Librarians] can be the Greyhound and Trailways of the information superhighway," he said.

You can make "home" computers available to the people who cannot have them in their own homes. Using them, people whom the educational system and the economic system have failed will be able to explore the bewildering array of programs that are available to them on the city, county, state and federal levels, and can be coached by software through the completion of the necessary applications, which they can then submit over the Net. That will not just give them a reason to learn to read; it will actually teach them computer skills that may permit them to land jobs. And in the process they will have learned to read. Think of it; at the same time that they see the need, they will be acquiring the skill. Furthermore, they will have acquired it in the library. Your library. It will no longer be an alien place, a place frequented by the educated elites whom they believe are their oppressors, but a known and friendly place.

Due to concerns regarding the development of a digital divide, federal government and corporate funding have provided a system of public Internet-access spaces throughout the United States. These free-access clusters are generally found in libraries and schools. By the late 1990s, many states began requiring that students pass a basic computing-proficiency examination before they are allowed to graduate from public schools, thus encouraging the teaching of networking skills.

THE HIGHWAY METAPHOR ROLLS ON
Following are twelve more illuminating statements made between 1990 and 1995 in which the highway metaphor comes into play.

Computers will become a truly useful part of our society only when they are linked by an infrastructure like the highway system and the electric power grid, creating a new kind of free market for information services. Imagine the United States without its highways. Our millions of cars, buses, and trucks would be driven in our own back yards and neighborhood parking lots, with occasional forays by the daring few along the uncharted, unpredictable, and treacherous dirt roads, full of unspeakable terrors. A ridiculous picture? Perhaps, but it is not far off the mark if we are talking not about transportation but about today's computers and the exchange of information among them. —MIT researcher and administrator **Michael Dertouzos,** in his 1991 article for *Technology Review* titled "Building the Information Marketplace"

It is possible that once computer networks become as commonplace as our national highway system, we will learn to treat them in much the same way. Rules of the road will emerge, and people will learn to respect them for their own safety and for the common good.—Technology writers **Katie Hafner and John Markoff,** in their 1991 book *Cyberpunk*

The nation's telecommunications giants are poised to spend billions of dollars during the next decade to revamp their electronic networks and to lay millions of miles of fiber-optic cable. Investment in the Information Superhighway could generate jobs and swell incomes for years. Indeed, just as economic growth in the 1980s was kicked higher by real estate investment, economic growth in the 1990s may be driven by the telecommunications revolution. . . . This investment increase could rival the rise in business spending on computers in the past several years [$22 billion]. It could rival federal spending on the new interstate highway system during the late 1950s and early 1960s [$38 billion]. —Technology writers **Christopher Farrell and Michael Mandel**, in a 1993 article for *Business Week* titled "What's Arriving on the Information Highway? Growth"

Because the Internet is a network of networks, no one group or person is in charge. . . . It's definitely out of control. . . . The role of capital as an editor is being removed. The metaphor of the highway fits. Like Jack Kerouac's "On the Road," from a tight little community out onto the wide-open road. Everybody's out there; it's not a small, elite system.—R.U. **Sirius** (real name, Ken Goffman), editor of the cyberpunk magazine *Mondo 2000,* in a 1993 article in *The Nation* titled "The Whole World Is Talking"

It's like trying to predict back in 1910 the impact of the automobile on society—the highway system, gasoline refineries, motels instead of hotels, new dating patterns, increased social mobility, commuting to work, the importance of the rubber industry, smog, drive-thru restaurants, mechanized warfare, and on and on. The net will bring more than quantitative changes, it will bring "qualitative" changes. Things that were impossible will now become inevitable.—Technologist **Larry Landwehr**, in a 1993 article in *Computer Underground Digest* titled "Some Comments on the London Times Supplemental Article"

I don't understand the hullabaloo. Can you imagine a more frivolous way to spend billions, hooking up people's homes so that kids can compete playing Marioworld? The Information Highway is a buzzword, created by public-relations people and folks at the White House who want to be seen as technology visionaries. To me, it doesn't mean anything.—Venture capitalist **Don Valentine**, quoted in a 1994 *Newsweek* magazine article titled "High Tech, High Risk"

Any future information network will help unhappy people secede, at least mentally, from institutions they do not like, much as the interstate highway system allowed the affluent to flee the cities for the suburbs and exurbs. Prescribing mobility, whether automotive or electronic, as an antidote to society's fragmentation is like recommending champagne as a hangover remedy. —Commentator **Edward Tenner**, in a 1994 essay for *Wilson Quarterly* titled "Learning from the Net"

Internet traffic started out like a group of cows. At first, there were just a few cows, and a farmer would let his neighbor's cows pass through his pasture free of charge. But now, there are more and more cows and even an occasional horse or elephant. . . . In order to accommodate the increased amount of traffic on the Internet, one farmer or a group of farmers will have to stand up and say, "I want to build a highway that bypasses this backroads pasture stuff." . . . But if we're going to build that highway, we have to be able to figure out how people can use it for cheap. I think we should make different classes of service available with different pay structures. What I think hasn't been explained to [U.S. Vice President] Al Gore is we can have one class of service that is paid for by the rich, and another one that is cheap and easy to get on. We can run them in parallel.—NetWorld and CyberCash founder **Daniel Lynch**, quoted in a 1994 article in *Telecommunications* magazine titled "An Interview with Interop founder Dan Lynch"

Far from acting as a general tool for freedom, automobiles disenfranchise relatively large segments of the middle class, the young, the elderly, and the rural from full participation in public life. How different would the public interpretation of computer/communication networks be if the guiding metaphor were that of the "information sidewalk"? Most importantly, a vital sidewalk life promotes feelings of community, of connections, of lives lived in common.—Graduate student **John Monberg,** in his 1994 *Computer-Mediated Communication* article titled "Welcome to the Emerald City! Please Ignore the Man Behind the Curtain"

The term "highway" . . . suggests that everyone is driving and following the same route. This network is more like a lot of country lanes where everyone can look at or do whatever his individual interests suggest. Another implication is that perhaps it should be built by the government, which I think would be a major mistake in most countries. But the real problem is that the metaphor emphasizes the infrastructure of the endeavor rather than its applications. . . . A different metaphor that I think comes closer to describing a lot of the activities that will take place is that of the ultimate market. . . . It will be where we social animals, sell, trade, invest, haggle, pick stuff up, argue, meet new people, and hang out.—Microsoft CEO **Bill Gates,** in his 1995 book *The Road Ahead*

The multi-billion-dollar investments in the developing new telecommunications landscape are driven by a simple, irresistibly tempting vision—the prospect of converting every home and workplace in the nation into a computerized electronic movie theater; shopping mall; video game arcade; business, information and financial center; and perhaps even gambling casino, run by remote control and open all day long, every day of the week. The information highway will not be a freeway but an automated, private toll road.—Former NBC and PBS executive **Lawrence Grossman,** in his 1995 book *The Electronic Republic*

Without a thoughtful universal-service policy, cyberspace could well end up as alien and cost-prohibitive to the general public as venturing out of town was during the reign of medieval highway robbers.—Progressive Networks founder **Rob Glaser,** in a 1995 essay for *Wired* magazine titled "Universal Service Does Matter"

**AND A BIT OF 1990s HUMOR FROM
THE ELECTRONIC FRONTIER FOUNDATION SITE**
The Top Ten Anagrams for "Information Superhighway":

10. Enormous, hairy pig with fan
 9. Hey, ignoramus—win profit? Ha!
 8. Oh-oh, wiring snafu: empty air
 7. When forming, utopia's hairy
 6. A rough whimper of insanity
 5. Oh, wormy infuriating phase
 4. Inspire humanity, who go far
 3. Waiting for any promise, huh?
 2. Hi-ho! Yow! I'm surfing Arpanet!

And the number one anagram for "Information Superhighway":

 1. New utopia? Horrifying sham

Knocking the Net

Some warn the Internet is naughty, anti-nature, and nefarious; even supporters see negatives

The explosion of networked digital communications in the 1990s led to an unprecedented level of scrutiny and criticism of the technology—never before had a new medium been so thoroughly and publicly debated. As is typical in the modern world, the majority of the voices spoke out in favor of new technologies tied to a new form of networked communications, but there were detractors. Heading up public opposition to passive acceptance of a networked world were Neil Postman, Paul Virilio, Sven Birkerts, Kirkpatrick Sale, and Clifford Stoll. Because it was still an early stage of the Internet's development, even the supporters and curious experimenters were finding some faults.

Technology writer Steven Levy illustrated several prime issues of concern in a 1995 article in *Newsweek* magazine titled "The Year of the Internet" in which he wrote: "The revolution is only just begun. It's already starting to overwhelm us, outstripping our capacity to cope, antiquating our laws, transforming our mores, reshuffling our economy, reordering our priorities, redefining our workplaces, putting our Constitution to the fire, shifting our concept of reality."

Self-proclaimed neo-Luddite Kirkpatrick Sale said human societies could be destroyed by networked communications technologies. "This time around the technology is even more complex and extensive [than that of the industrial revolution], and its impact even more pervasive and dislocating, touching greater populations with greater speed and at greater scales," he said in a 1995 *Wired* magazine article titled "Return of the Luddites." "[If what's in store for us] is an information age with information highways and information supermarkets, then it is the computer and those who feed and handle it who reign supreme: in the country of the sighted, the all-seeing one is king. Control of information is control of power."

Physicist and network user/critic Clifford Stoll wrote in his 1995 book *Silicon Snake Oil:* "The medium is being oversold, our expectations have become bloated, and there's damned little critical discussion of the implications of an online world."

FROM THE FIRST, TECHNOLOGY HAS HAD ITS OPPONENTS

The term *technology* was initially used in Europe in the late 1700s to refer to the contents of books that transferred a craftsman's "knowledge of the arts" from one generation to the next. Karl Marx, in his *Economic Manuscripts of 1861–63*, was one of the first to document this origin of the term, which comes from the Greek *techni,* meaning art or skill, and *ology,* or "the study of."

Current use of the word began its evolution in 1829, when Jacob Bigelow, Harvard's Rumsfeld Professor of Physical and Mathematical Sciences as Applied to the Useful Arts, assembled his lecture notes into a book he titled *Elements of Technology.* The book's contents indicated that the word *technology* included in its meaning materials, equipment, structures, and processes, and the knowledge and skills to make use of them.

Today we define technology as the innovation and use of tools, machines, materials, and/or processes to accomplish something. The word is used to describe such innovations and uses dating back to the earliest days of humankind, thus the wheel and the hammer are just as much technologies as is the computer.

GOOD "KING LUDD" INSPIRED RESISTANCE MOVEMENTS

Those who oppose some or all new technologies have been referred to as neo-Luddites or technophobes. The root of the "Luddite" label, attached to those who express opposition, radical dislike, or pure hatred for new technologies, had its origins in Leicestershire, England, in the late 1770s. A man named Ned

Ludd had destroyed two hosiery-knitting machines in a fit of rage. Despite the fact that this type of protest had been taking place for at least a century, it became common thereafter to use the phrase, "King Ludd must have been here" whenever machines of any sort were found to be smashed.

Between 1811 and 1816, when masked attackers began battling poor working conditions suffered by textile workers, they swore their allegiance to "King Ludd." Officials of the time turned the name around to their own use, referring to those "unenlightened" individuals who wished to obstruct technological progress as Luddites. Because wool and cotton mills had been destroyed by such acts of protest, the British government made "machine breaking" a capital crime, and 17 men were executed for it in 1813. Romantics of the time identified with the attackers' cause. Lord Byron wrote the poem "Song for the Luddites" in 1816. The first stanza reads, "As the Liberty lads o'er the sea / Bought their freedom, and cheaply, with blood, / So we, boys, we / Will die fighting, or live free, / And down with all kings but King Ludd!"

A short time later, Mary Shelley, daughter of leading feminist Mary Wollstonecraft and wife of the poet Percy Shelley, wrote the 1818 horror-fiction classic *Frankenstein*, a cautionary tale about the pursuit of scientific advancement without regard to moral consequences. In a telling passage of the novel, the monster says to Dr. Victor Frankenstein, "You are my creator, but I am your master." In the scope of today's entertainment and technology, science-fiction films such as *Blade Runner, The Matrix*, and *I, Robot* echo the warnings Shelley put forth in *Frankenstein*.

CRIMINAL KACZYNSKI MADE SOME VALID POINTS

Theodore Kaczynski could be considered the Ludd of more modern times. Concerned about the changes brought by technology, he engaged in an 18-year mail-bombing campaign, killing three and wounding 29 between 1978 and 1994. He became known as the Unabomber (from the FBI reference to him as the "university and airline bomber"), and his anonymous attacks were often spread years apart.

A brilliant man, he graduated from Harvard at the age of 16, earned a doctorate in mathematics from the University of Michigan, and worked in complex analysis before retreating to a remote shack in Montana. It was from this location that Kaczynski worked and sent his bombs as a form of protest against technology to university professors, airlines, computer stores, an advertising executive, and a forestry association president.

In 1995, he mailed letters to some former victims, requesting that a statement outlining his argument be printed in a major newspaper. The statement, "Industrial Society and Its Future," also known as The Unabomber Manifesto, was printed, after some argument, in the *New York Times* and *Washington Post* in September 1995. The publishers did not want to give in to a murderer's demands, but it was thought that if the piece was published it might be easier for the public to help identify the bomber.

In his written statement, Kaczynski argued that technological progress is negative and should be stopped so people can live simpler, more natural lives. He wrote:

> Technology advances with great rapidity and threatens freedom at many different points at the same time (crowding, rules and regulations, increasing dependence of individuals on large organizations, propaganda and other psychological techniques, genetic engineering, invasion of privacy through surveillance devices and computers, etc.) To hold back any ONE of the threats to freedom would require a long, difficult social struggle. . . . First, we must work to heighten the social stresses within the system so as to increase the likelihood that it will break down or be weakened sufficiently so that a revolution against it becomes possible. Second, it is necessary to develop and propagate an ideology that opposes technology and the industrial society if and when the system becomes sufficiently weakened.

Kaczynski's brother David recognized the writing style and notified the authorities, and officers arrested the Unabomber April 3, 1996, at his cabin near Lincoln, Montana. He avoided the death penalty by pleading guilty January 22, 1998, and was sentenced to life in prison with no parole.

PROTESTORS INCLUDE SOCRATES, ROUSSEAU, AND THOREAU

Ludd and Kaczynski are members of a select group of historically significant figures who have opposed technological progress. In *Phaedrus,* Socrates expressed his doubts about the development and use of the written word, predicting that it would weaken people's memories, alter the practice of education, and make public those things best left private. Every communications technology developed in the 2,500 years since he said this has rightfully been met with the same questioning scrutiny.

Jean-Jacques Rousseau first expressed the point of view now labeled *Romanticism* in an essay published in 1749. In it, he questioned the scientific progress of his time, suggesting that such so-called advances would lead to atheism, materialism, and the diminishing of the human spirit. Romanticism became a worldwide movement, and by the 1800s, most knowledgeable people divided themselves into two camps: the Romantics, who believed in the merits of the arts, traditional crafts, poetry, and the like, opposing industrialization, new technologies, and machines; and the Progressives, who saw developments in science and technology as vital advances for the good of society.

A leading Transcendentalist of the 19th century, Henry David Thoreau, remarked in his 1854 book *Walden*, "We are in great haste to construct a magnetic telegraph from Maine to Texas; but Maine and Texas, it may be, have nothing important to communicate. . . . We are eager to tunnel under the Atlantic and bring the old world some weeks nearer to the new; but perchance the first news that will leak through into the broad, flapping American ear will be that the Princess Adelaide has the whooping cough."

UTOPIANS SEE GOOD IN TECHNOLOGY;
TECHNOREALISTS SEE GOOD AND BAD

Despite the occasional heavy-hitting detractors, historians say most Americans embraced the technological advances that have reshaped the world over the past 250 years. In fact, by the late 1800s supporters of science and technology began describing their heroes in Romantic terms, celebrating the work of such inventors as Morse, Edison, and Bell in poetry and popular music. Fictional accounts of a utopian society laced with beneficial technologies were popular in the latter 19th century. Examples include Edward Bellamy's *Looking Backward: 2000–1887*, written in 1888, and King Camp Gillette's *The Human Drift*, written in 1894. Their projected utopias would be an idealized combination of the arts and subjectivity of the Romantics and the sciences and objectivity of the Progressives. Their perfect world saw technology bringing prosperity, leisure, comfort, creativity, free expression, and happiness.

Today's technorealists work to expand the middle ground between the utopians and the neo-Luddites. They see the odds for the development of such a complete utopia to be slim. Complicated technologies are born out of trial and error, with plenty of problems along the way. Such world-changing developments also bring jarring changes to the social structures to which they're introduced. Computers, computer networks, and the people who use them have been ridiculed, reviled, and picked apart by concerned critics.

Technology writer and Internet enthusiast Bruce Sterling said in a 1991 interview for *Compute!* "[People in general say they] are very afraid of computer hackers, but I think mostly they are afraid of computers. . . . A computer hacker puts the face on the menace that is represented by computers. I mean, I am afraid of computers. . . . I am not computerphobic; I'm rationally afraid of computers. Computers are a challenge and a threat, and they're changing our society in ways that we can't control and don't understand. They're not to be trusted."

NEIL POSTMAN WAS A TRUE ROMANTIC

Communications technology critic Neil Postman expressed the Romantic attitude when he said in his 1990 speech "Informing Ourselves to Death," delivered before the German Informatics Society:

> In a world populated by people who believe that through more and more information, paradise is attainable, the computer scientist is king. But I maintain that all of this is a monumental and dangerous waste of human talent and energy. Imagine what might be accomplished if this talent and energy were turned to philosophy, to theology, to the arts, to imaginative literature, or to education? Who knows what we could learn from such people—perhaps why there are wars, and hunger, and homelessness, and mental illness, and anger.

Postman went on, enumerating some of the evils accompanying a fully networked society:

> To what extent has computer technology been an advantage to the masses of people? . . . These people have had their private matters made more accessible to powerful institutions. They are more easily tracked and controlled; they are subjected to more examinations and are increasingly mystified by the decisions made about them. They are more often reduced to mere numerical objects. They are being buried by junk mail. They are easy targets for advertising agencies and political organizations. The schools teach their children to operate computerized systems instead of teaching things that are more valuable to children.

The critic went on to conclude that everyone is being sold a bill of goods:

> It is to be expected that the winners . . . will encourage the losers to be enthusiastic about computer technology . . . they tell them that their lives will be con-

ducted more efficiently, discreetly neglecting to say from whose point of view or what might be the costs of such efficiency.... The computer is, in a sense, a magnificent toy that distracts us from facing what we most needed to confront— spiritual emptiness, knowledge of ourselves, usable conceptions of the past and future.... Through the computer, the heralds say, we will make education better, religion better, politics better, our minds better—best of all, ourselves better. This is, of course, nonsense.

The speech was immediately embraced by people who saw some truth in its words, whether or not they used technology themselves. The complete text or segments thereof have been reprinted literally thousands of times on the Internet. Postman's anti-tech message has been and is being carried to the masses by the very thing he condemned.

VIRILIO WARNS AGAINST "VIRTUALIZATION"; DEBRAY PREDICTS CONFLICT

French anti-technology theoretician Paul Virilio proposed in a 1995 interview with *Wired* magazine titled "Speed and Information" that in the 1990s Internet age "the very word 'globalization' is a fake. There is no such thing as globalization, there is only virtualization. What is being effectively globalized by instantaneity is time. Everything now happens within the perspective of real time: henceforth we are deemed to live in a 'one-time-system.' For the first time, history is going to unfold within a one-time-system: global time. Up to now, history has taken place within local times, local frames, regions, and nations. But now, in a certain way, globalization and virtualization are inaugurating a global time that prefigures a new form of tyranny."

Virilio said he mourned the loss of the "local" in the world.

In the very near future, our history will happen in universal time, itself the outcome of instantaneity—and there only.... Thus we see on one side real time superseding real space. A phenomenon that is making both distances and surfaces irrelevant in favor of the time-span, and an extremely short time-span at that. . . . There is talk of substituting the term "global" by "glocal," a concatenation of the words local and global. This emerges from the idea that the local has, by definition, become global, and the global, local. Such a deconstruction of the relationship with the world is not without consequences for the relationship among the citizens themselves.

And fellow French theoretician Regis Debray, self-proclaimed creator of a discipline he called "mediology," predicted in 1995 *Wired* interview titled "Revolution in the Revolution" that technological advances would cause imbalances in the world that would lead to conflict—exactly the conflict that followed in Africa and the Middle East. "Each technical step forward means a compensating step backward in our mind-sets," said Debray. "Islamic fundamentalists don't come from the traditional universities deeply rooted in a literary educational system; they graduate from engineering schools and technical colleges. Last century, some futurologists foresaw the end of national wars under the influence of spreading railroad lines and electrical telegraphy; others believed that industrialization would wipe out religious superstition. In fact, an imbalance in technologies tends to provoke a corresponding refocusing on ethnic values."

Debray went on, speaking out for a concentrated effort to retain media diversity in an effort to retain peace and individuality of cultures:

> I think we should negotiate a contract for mediodiversity in a mediosphere that is continually threatened with increasing uniformity of content because of the spread of global networks. . . . By transforming three-quarters of the world into a cultural proletariat, you will make people of this class into more determined rebels in the 21st century. Far more determined, in fact, than the economic proletariat has been in the 20th century . . . machines will never be able to give the thinking process a model of thought itself, since machines are not mortal. What gives humans access to the symbolic domain of value and meaning is the fact that we die.

During the Internet's "awe" stage in the early 1990s, the voices of neo-Luddites chimed in, and many of the network's biggest supporters, including Mitch Kapor, Howard Rheingold, Marc Andreessen, and Bill Gates, were technorealist enough to recognize the importance of acknowledging the likely problems to come packaged with the positive aspects of this new way to communicate.

Following are some of the concerned comments made between the years 1991 and 1995 by people engaged in the debate over the impact of the technology of networked communications.

WORDS OF WARNING FROM 1991

People will become socially immature. Virtual realities do what people want them to do, and that's not the way the real world works.—**Thomas Furness, a** professor at the University of Washington, quoted in a *U.S. News & World Report* article titled "Are New Realities More or Less Real?"

Computers, like other technological creations, create undesirable as well as desirable possibilities. We now have a greater capacity to track and monitor individuals without their knowledge, to develop more heinous weapons systems, to eliminate the need for human contact in many activities.—**Deborah Johnson,** author of *Computer Ethics,* quoted in a *National Forum* article titled "Computers and Ethics"

Technologies can, presently most certainly are, and thus probably will be, used to decrease democracy, sometimes perhaps while actually increasing "participation." While I have no doubt that these technologies will increase participation, without our active imagination, design, and effort, I am certain that they will not be used to make democracy possible but rather merely to concentrate even more the power of the very few who control the present. Indeed, even with our active imagination, design, and effort, these communication technologies might be used to enhance the power of the already powerful. . . . Should this be called "quantum fascism?"—Futurist **Jim Dator,** in a speech he made before the Communications Working Group at the Conference of the World Futures Studies Federation

CONCERNED CRITICISM FROM 1993
In the worst case, we could wind up with networks that have the principal effect of fostering addiction to a new generation of electronic narcotics. . . . Their principal themes revolving around instant gratification through sex, violence, or sexual violence; their uses and content determined by mega-corporations pushing mindless consumption of things we don't need and aren't good for us. . . . Purveyors of network services could simply decide that a business strategy that encourages the widest variety of content sources and originators will dramatically increase network usage. A few pennies per transaction will eventually add up to billions of dollars in revenue.—EFF co-founder **Mitchell Kapor,** in an article for *Wired* headlined "Where Is the Digital Highway Really Heading?"

Information has now become a form of garbage. We don't know what to do with it, have no control over it, don't know how to get rid of it. In the face of this, we propose to spend billions on a super-information network. To do what? Instead of 60 TV channels, we'll have 500, maybe a thousand. We'll have access to more entertainment, more sports, more commercials, more news—faster, more conveniently, in more diverse forms. We will, in other words, flood our lives with that from which we are already drowning.—**Neil Postman,** quoted in a *New York Times* article headlined "We Are the Wired"

With millions of users pounding away on their keyboards, it may turn out that everybody is talking and nobody is really listening. By the same token, connecting with one and all in the electronic ether could leave people more disconnected than ever before, as the necessity of face-to-face contact diminishes. If a troubled or shy office worker easily finds solace and approval on the networks, will she be less inclined to seek out friends on the job? —Writer **Laurie Hays,** in a article for *The Wall Street Journal* titled "A New World: Amid All the Talk About the Wonders of Networks, Some Nagging Social Questions Arise"

People are becoming addicted to all this stuff. It's like television. Look at the education fall-off since television began. Why? Kids are vegging out. People are becoming isolated from the outside world, literally and figuratively. They're dealing with pieces of equipment, rather than each other.—Futurist **Gerald Celente,** quoted in a *Boston Globe* article headlined "Online All the Time"

Public life will have disappeared because we did not see, in time to reverse the process, that our dazzling technologies were privatizing almost all social activities. It became first possible, then necessary, to vote at home, shop at home, listen to music at home, see movies at home. We replaced our libraries with interactive videotext, which even encouraged individuals to change the ending of stories if they didn't like them, so that there no longer existed common literature. We replaced schools with home computers and television. We replaced meeting friends with the video telephone and electronic mail. We replaced visits to far away places with virtual reality. We became afraid of real people and eventually forgot how to behave in public places, which had become occupied almost entirely by criminals. The rest of us had no need to be with each other.—**Neil Postman,** in an essay he wrote for *U.S. News & World Report* titled "Beyond 1993"

ISSUES DISCUSSED IN 1994

There will be no "dead air" on our new conduits. This doesn't mean that content is king though, for if history is any guide, most of it will be fungible commodity properties such as game shows or old movies today, or utter trash. . . . Now the "vast wasteland" . . . will be supplanted by a vaster wasteland brimming with utterly new forms of interactive cyberdreck.—Futurist **Paul Saffo,** in a *Wired* essay titled "It's the Context, Stupid"

In the future, as the information superhighway looks more like an electronic marketplace, "digital cash" might be vulnerable to theft. "Burglars" might be able to break into a computer and download cash, and "muggers" might be able to rob intelligent agents that have been sent out on the network with cash to purchase information goods.—Network security expert **Dorothy Denning,** in an article for the *Journal of Criminal Justice Education* titled "Crime and Crypto on the Information Superhighway"

Information technologies pose three types of problems; they intrude on personal privacy; they offer the means for institutions to control their clients; and they encourage practices that threaten certain democratic values.—**Dorothy Nelkin,** in a *National Forum* article headlined "Information Technologies Could Threaten Privacy, Freedom, and Democracy"

Computer viruses and worms are maliciously constructed agents—fanning out, like Fagin's boys, to cause trouble. Will there be a criminal underclass? . . . Since agents are easy to reproduce, cyberspace may be flooded with billions of them; how will population be controlled? . . . Even if our agents turn out to be very smart, and always perform impeccably, will we ever fully trust them? . . . We will want our agents to be as smart as possible in order to do our bidding most effectively, but the more intelligent they are, the more we will have to worry about losing control and the agents taking over. . . . The burgeoning, increasingly indispensable, programmed proletariats of cyberspace cities now live invisibly on disk drives.—MIT dean and professor **William Mitchell,** in his book *City of Bits*

New information technologies can easily be turned to malign ends. Through advertising and other means, they have been used not only to exploit our hearts' desires but to manufacture new ones. Along with the specter of greater government control over citizens' lives that becomes possible with the new information technologies, this "commodification of desire" must be considered one of the darker prospects of the NII. . . . With the NII, it seems likely that the machines will grow stronger, as will marketers and governments.—Tom Maddox, in an essay for *Wilson Quarterly* titled "The Cultural Consequences of the Information Superhighway"

The fact that the information highway is also a global network makes it easier for corporations to acquire the use of highly skilled but less expensive labor in overseas markets. . . . This high-speed data network facilitates a process of pre-employment screening that includes not only credit ratings and criminal records, but also information about health and personal habits. . . . This same superhighway will facilitate the management of labor by means of work assignment and by monitoring and surveillance techniques that build mistrust and contribute . . . to a level of stress that threatens both health and productivity.—Communications researcher Oscar Gandy, in his *National Forum* article "The Information Superhighway as the Yellow Brick Road"

What the hypemeisters don't say or don't realize is that this frontier metaphor [attached to virtual communities] deceives us. It conjures up Americana images of the individual lighting out for the territories, independent and hopeful, to make a life. But what is hidden by the metaphor is the cybernaut immersed in virtual worlds, neither self-reliant nor liberated, but utterly dependent for existence on technology created, provided, and sustained by others, living the isolated life of the placeless domesticate.—Researcher/commentator Stephen Doheny-Farina, in an essay in *Computer-Mediated Communication* titled "The Last Link: Cybernauts in the Electronic Frontier: Pets on a Leash"

POINTS TO PONDER FROM 1995

Identified by and targeted for our product consumption, we will find ourselves receiving more personalized mail from products than from people. They will know us, and they will manipulate us. We will end up hating the Internet, and ourselves.—Justin Hall, consultant and former *Wired* staffer, in an essay on his personal website titled "I Am Not My Habits: On Our Guard Against Targeted Advertising"

Unfortunately Tim Berners-Lee forgot to make an expiry date compulsory. It means that any information can just be left and forgotten. It could stay on the network until it is five years out of date.—**Brian Carpenter,** head of communications systems at CERN, in an article run by the Associated Press headlined "Information Highway Has Litter Problem"

Internet hustlers invade our communities with computers. . . . The key ingredient of their silicon snake oil is a technocratic belief that computers and networks will make a better society. [But] the most important interactions in life happen between people, not between computers. . . . One of the more pernicious myths of the online world is that of a literary revival. Instead of an Internet-inspired renaissance, mediocre writing and poorly thought-out arguments roll into my modem. E-mail and postings to network newsgroups are frequently ungrammatical, misspelled and poorly organized. . . . What does computer literacy mean to a child who can't read at grade level and can't interpret what she reads?—**Clifford Stoll,** in excerpts from his book *Silicon Snake Oil*

I urge people to take a clear-headed look at what is in front of them, and not to feel guilty if they reject something, and to be able to say, with a rational explanation, This is wrong, I will not myself buy into it. . . . There are two moral judgments against computers. One is that computerization enables the large forces of our civilization to operate more swiftly and efficiently in their pernicious goals of making money and producing things. . . . And secondly, in the course of using these, these forces are destroying nature with more speed and efficiency than ever before.—**Kirkpatrick Sale,** in a story in *Wired* magazine headlined "Interview with the Luddite"

Few aspects of daily life require computers, digital networks, or massive connectivity. They're irrelevant to cooking, driving, visiting, negotiating, eating, hiking, dancing, and gossiping. You don't need a keyboard to bake bread, play touch football, piece a quilt, build a stone wall, recite a poem or say a prayer. . . . We're told that anyone without a modem is an inept bumpkin, hopelessly behind the times or afraid of the march of technology. Don't buy it, or the cyberbullies will bury us all. . . . Life in the real world is far more interesting, far more important, far richer, than anything you'll ever find on a computer screen. . . . Every hour that you're behind the keyboard is 60 minutes that you're not doing something else.—**Clifford Stoll,** in his book *Silicon Snake Oil*

Screen-based technologies (such as TV and computer monitors) are prone to induce democratically unpromising psychopathologies, ranging from escapism to passivity, obsession, confusing watching with doing, withdrawal from other forms of social engagement, or distancing from moral consequences. —Loka Institute communications researcher **Richard Sclove,** in his book *Democracy and Technology*

[Even the family will change.] What happens when families come back together because work is done at home? What neuroses will that expose? —*Wired* magazine publisher **Louis Rossetto,** in a *London Guardian* interview headlined "Inter Next World"

Searching for an easy fix, we are eager to believe that the Internet will provide an effective substitute for face-to-face interaction. But the move toward virtuality tends to skew our experience of the real in several ways. One way is to make denatured and artificial experiences seem real. Let's call it the Disneyland effect. . . . Another effect of stimulation, which I'll call the artificial crocodile effect, makes the fake seem more compelling than the real. . . . A third effect is that a virtual experienced may be so compelling that we believe that within it we've achieved more than we have. . . . To the question, "Why must virtuality and real life compete—why can't we have both?" the answer is of course that we will have both. The more important question is, "How can we get the best of both?"— Researcher **Sherry Turkle,** in her book *Life on the Screen*

I'm human, a social animal. I'm not a god, I'm a hairless chimp with a messianic complex and a mouse. I need human contact and simple pleasures. I need to eat, poop and see people smile. . . . But instead of returning to the basics, I, like many of us, am spending more and more of my time with my face bathed in monitorglow, getting my fix of digital junk. Won't someone please unwire me before it's too late?—**Christopher Scheer,** in his essay for *The Nation* headlined "The Pursuit of Techno-Happiness"

Anyone who's directed away from social interactions has a head start on turning out weird. . . . Computers teach us to withdraw, to retreat into the warm comfort of their false reality. Why are both drug addicts and computer afi-cionados both called users?—**Clifford Stoll,** in his book *Silicon Snake Oil*

To its critics, teledemocracy conjures the image of alienated, silent voters, sit-ting alone inside an electronic cocoon. . . . These modern-day robotic voters absorb virtually all their information about the outside world electronically —through CNN-style live coverage, tabloid-news shows, and sensational newsmagazines; and, increasingly, at programmed computer terminals. The wired public then feeds back its ill-formed, unsophisticated, unmediated opinion instantaneously, without deliberation, following simple on-screen in-structions to press Y for yes, or N for no. . . . But is that actually the way elec-tronic democracy will operate, or needs to operate in the future?—Former NBC and PBS TV executive **Lawrence Grossman,** in his book *The Electronic Republic*

Culturally, we may lose touch with the mainstream. There may be no more mainstream. Politically, we may find ourselves detached from the dialogue. Right now, the president can talk over all three networks to us at the same time. With 500 video channels, that won't be so easy.—Author **Les Brown,** quoted in a *Seattle Post-Intelligencer* article headlined "Scholars Try to Mea-sure the Impact"

Future network users will suffer from infoglut; the volume of resources avail-able on the Internet will be completely overwhelming. There will be lots of gold, but more fool's gold, tailings, and junk. If you search for evolution, you may get Darwin's Origin of Species, fourth grade essays, creationist dogma, and erudite scientific papers. Mediocre and dreadful resources will be on the networks in far greater quantity than excellent ones, as always.—Education technology expert **Bob Tinker,** in a white paper titled "The Future of Net-working Technologies for Learning"

Computers don't make any . . . old free-expression problems go away; on the contrary, they intensify them, and they introduce a bunch of new problems. Problems like software piracy. Encryption. Wire fraud. Interstate transporta-tion of stolen digital property. Free expression on privately owned networks. So-called "data-mining" to invade personal privacy. Employers spying on employee e-mail. Intellectual rights over electronic publications. Computer search-and-seizure practice. Legal liability for network crashes. Computer intrusion. And on and on and on. These are real problems. They're out there. They're out there now. In the future, they're only going to get worse. And there's going to be a bunch of new problems that nobody's even imagined. —**Bruce Sterling,** in a speech he made at the 1995 High Technology Crime Investigation Association Conference

Overwhelmed by data and distracted by fantasy, our attention span could become reduced to the content-poor shards of an MTV video. We risk becoming mentally poorer, mistaking data for knowledge, distanced from wisdom and blind to both beauty and the terrors of intangible, felt existence.—British philosopher **Michael Heim,** quoted in a *Seattle Post-Intelligencer* article headlined "Scholars Try to Measure the Impact"

There are lots of potential negatives for individuals. It's going to mean wrenching changes, I think. The technology itself: no negatives. The impact on people: it depends. It's going to continue to change the way the work force is put together, the activities that people undertake, the amount of change people have to be undergoing to stay current. It's going to challenge the education system in order to keep up. It's going to have a whole series of impacts like that. The sheer rate of change, assuming it continues to increase, is going to start to get pretty severe.—Netscape founder **Marc Andreessen,** in a video interview for the Smithsonian Institution Oral and Video Histories Series

The availability of virtually free communications and computing will alter the relationships of nations, and of socioeconomic groups within nations. The power and versatility of digital technology will raise new concerns about individual privacy, commercial confidentiality, and national security. There are, moreover, equity issues that will have to be addressed. The information society should serve all of its citizens. . . . Men and women are worried that their jobs will become obsolete, that they won't be able to adapt to new ways of working, that their children will get into industries that will cease to exist, or that economic upheaval will create wholesale unemployment, especially among older workers. These are legitimate concerns. Entire professions and industries will fade. But new ones will flourish. This will be happening over the next two or three decades.—**Bill Gates,** in his book *The Road Ahead*

The next decade will see cases of intellectual-property abuse and invasion of our privacy. We will experience digital vandalism, software piracy, and data thievery. Worst of all, we will witness the loss of many jobs to wholly automated systems, which will soon change the white-collar workplace to the same degree that it has already transformed the factory floor. . . . The radical transformation of the nature of our job markets, as we work less with atoms and more with bits, will happen at just about the same time the 2 billion-strong labor force of India and China starts to come on-line (literally). A self-employed software designer in Peoria will be competing with his or her counterpart in Pohang.—**Nicholas Negroponte,** in the epilogue to his book *Being Digital*

Saddam, O.J., and the Unabomber

Internet developments are tied to the news events and popular culture of the 1990s

An examination of America's pop culture, fads, and news events during radio's boom stage in the 1920s, TV's takeoff in the 1950s, and the introduction of the Internet in the 1990s can be instructive. As 20th-century communications technologies emerged over these decades, the personalities, politics, and policies of the times were shaped by and reflected in them. It's easy to see the similarities in the times and their patterns of growth.

As communications drew the nation closer over the course of the 20th century, and media outlets multiplied seemingly minute by minute, the number of recognizable personalities in the popular culture exploded. The constellation of famous folk recognized by the average American in the 1920s pales in comparison to the total number of such stars of the sports, entertainment, political, and corporate world of the 1990s.

As the communications forms began to mature between 1900 and 1999, the country's population numbers grew larger, paychecks and the average person's standard of living got fatter, and consumerism—thanks at least in part to the commercial interests propagated by modern media—drove the U.S.

economy to greater and greater heights. In each decade, corporate greed was revealed in one or more national scandals.

At the same time, the world became smaller and more dangerous. The end of the 1920s brought Adolph Hitler's Third Reich of the 1930s. The conclusion of the boom years of the 1950s brought the gut-wrenching 1960s: civil unrest, assassinations of major political figures, and the war in Vietnam. The end of the 1990s led into years of heightened religious and ethnic conflicts in many nations, including the wars in Afghanistan, Sudan, and Iraq.

In the beginning years of each decade in which a new technology began to boom, the American economy was strong and people were optimistic. In the closing years of each of these decades, the economy weakened or crashed and people clung to the hope that the good times would soon return.

TAKING A CLOSER LOOK AT THE "ROARING TWENTIES"

During the first big decade of radio—the post–World War I 1920s—the population of the United States was about 107 million, and life expectancy was about 54 years. In 1920, there were 9 million automobiles and by 1928 about 26 million cars were in use. There were 387,000 miles of paved road. The average worker's salary was $1,236.

Radio fans heard vivid descriptions of Babe Ruth's mammoth home runs and Charles Lindbergh's completion of the first transatlantic flight. The presidents were Warren G. Harding, Calvin Coolidge, and Herbert Hoover. In the 1920s, the 19th Amendment to the U.S. Constitution gave women the right to vote. Henry Ford became an automobile mogul, taking 50,000 orders for his Model A in 1927. Amelia Earhart and Annie Oakley were setting records. Gangster Al Capone was a newsmaker, and Prohibition drove people to socialize in speakeasies, drinking bootlegged alcohol.

Clarence Darrow and William Jennings Bryan debated evolution in a Tennessee courtroom in 1925, the same year that 40,000 Ku Klux Klan members marched on Washington. Mahatma Gandhi was imprisoned in 1922 for civil disobedience; King Tut's tomb was found; Mickey Mouse was introduced; Albert Einstein won a Nobel Prize. The Nazi Party had its first conference in 1923 and Adolph Hitler published *Mein Kampf* in 1925. Al Jolson starred in the first "talking picture," the film *The Jazz Singer,* in 1927.

Corporate greed and corruption was exposed in the Teapot Dome scandal of 1923, involving the leasing of naval oil reserves to private companies. In the middle of radio's "awe-stage" decade, on November 11, 1924, Wall Street broke a record when 2,226,226 shares were traded. At the end of radio's big decade, when the stock market crash came on Black Thursday, October 24, 1929, 16 million shares were traded and $30 billion vanished into thin air. Radio broadcasters shared all of the bad news, trying to put it into perspective. That and other economic woes led into the Depression of the 1930s.

TAKING A LOOK AT THE BOOMING 1950s

During the first big decade of television—the post–World War II 1950s—the population of the United States was about 150 million and life expectancy was 71 years for women and 66 years for men. The average annual salary for an American worker in 1950 was $2,992. It was a time of prosperity despite the fact that the United States was again at war. North Korea had invaded South Korea, and the United States sent troops to fight during the early 1950s as part of a United Nations effort to aid the South. Harry Truman and Dwight Eisenhower were the presidents. The U.S. government aided in the overthrow of Iran's government in 1953, and Guatemala's government in 1954.

Nikita Khrushchev, premier of the expansion-minded, communist-run Soviet Union, told Western ambassadors visiting Moscow in 1956: "History is on our side. We will bury you!" People began buying plans for bomb shelters, building the shelters, and preparing for civil-defense procedures thought practical in case of a nuclear attack. The Red Scare, Sputnik, and the Cold War emerged from the rise of communism and the continuing expansion of the Soviet Bloc.

In 1958, Eisenhower appointed MIT president James Killian as presidential assistant for science and created the Advanced Research Projects Agency (ARPA—the program from which the Internet evolved) to develop technologies to bolster the U.S. defense system and economy. In 1959, the seven original Mercury astronauts were selected by NASA, and the United States had launched its first Explorer satellite.

Television viewers watched Hank Aaron hit home runs. They saw Martin Luther King Jr. lead a transportation boycott in Montgomery, Alabama, in 1955; heard about a Supreme Court decision that required school integration; and witnessed violent demonstrations against integration in Little Rock, Arkansas.

Nearly 7 million cars were sold in 1950 alone. Construction of the interstate highway system was signed into law in 1956, and the first cars with seat belts came out the same year. Thanks to newfound mobility and the decade's prosperity and building boom, the first shopping malls were built and drive-in movie theaters were popular. By 1959, a total of 1.25 million Americans had died in motor vehicle accidents, more than had died in war; highways and fast cars had ushered in death as well as prosperity. At the end of the decade the United States was suffering its worst recession since the end of World War II. More than 5 million people were unemployed and 56 percent of new businesses failed.

THE GO-GO BUZZ OF THE NET-CRAZY 1990s

During the first big decade of the Internet—the post–Cold War 1990s—the population of the United States was about 280 million, life expectancy was 79 years for women and 73 years for men. The average annual salary in 1998 was $26,412. There were 2.4 million miles of paved road and 130 million motor vehicles, and as many as 49,000 people died each year in auto accidents. There were so many new fast-food chains opening up that the National Association of Uniform Manufacturers and Designers estimated the value of the employee-apparel industry rose to at least $6 billion. Not coincidentally, obesity was at a record high. Bill Gates became a computer mogul as Microsoft's software sales reached $1 billion in 1990. Worldwide, more than 100 million computers were in use. The presidents during this decade were George H.W. Bush and Bill Clinton.

People could see Barry Bonds, Sammy Sosa, and Mark McGwire hitting home runs on the Internet. They could go online and become actively involved in campaigning for the welfare-reform and Americans-with-disabilities legislation passed during this decade. Americans watched on live television and shared their opinions on the Internet in 1994, as former football star and movie actor O.J. Simpson tried to elude police, driving his white Bronco along California's freeways, and they reacted online to his 1995 trial, broadcast on live television, as he was acquitted of the murders of his ex-wife Nicole and her friend Ron Goldman.

In 1992, rioting and looting resulted after an all-white jury in suburban Simi Valley, California, acquitted four white Los Angeles police officers on all but one charge stemming from the March 1991 beating of black motorist Rodney King; more than a hundred arson fires were set during the riots, and a dozen people died.

After the fall of the Berlin Wall in late 1989—symbolic of the end of the Cold War—there were doubts and hopes about how the region that was formerly the

Soviet Union would reinvent itself. Most of the 900,000 tons of concrete left when the wall was pulverized were used in road construction. Boris Yeltsin was elected president of Russia in 1990.

U.S. troops were involved in a number of worldwide police actions: they fought in the six-week Gulf War I in 1991, after Saddam Hussein invaded Kuwait; went to Somalia in 1993 to try to aid the starving people during the regime of warlord General Adid; overthrew the military dictatorship in Haiti in 1994; were deployed to Bosnia in 1996 as part of a NATO peacekeeping force; and joined NATO in 1999 air strikes meant to halt the Yugoslavian government's policy of ethnic cleansing in Kosovo.

Heinous acts of domestic and international terror were committed. In 1993, Islamic rebels detonated a bomb in the garage beneath the World Trade Center in New York, killing six and injuring more than 1,000. This first overt act of such terror committed by foreign perpetrators was a shock to the nation. In 1995, people were even more stunned when Timothy McVeigh and a group of U.S. citizens used a truck bomb to destroy the federal building in Oklahoma City, killing 168 men, women, and children. The attack came on the anniversary of the 1993 Branch Davidian Massacre, in which as many as 77 people led by David Koresh were killed in a fire resulting from a standoff with agents from the Bureau of Alcohol, Tobacco, and Firearms. In 1999, Eric Harris and Dylan Klebold, teen fans of violent films and video games who had revealed their violent tendencies on the Internet, attacked their teachers and fellow students at Columbine High School in Littleton, Colorado, killing 15 (including themselves) and wounding 23.

$1.3 TRILLION SPENT ON INFORMATION TECHNOLOGY

According to the Institute for U.S. Policy Studies, the U.S. economy boomed in the 1990s, as corporate profits jumped 108 percent over the decade, supporting an S&P increase of 224 percent. In the first nine years, pay for chief executive officers went up 443 percent (from an average total compensation of $2 million in 1990 to $10.6 million in 1998), and workers' pay was up 28 percent (with no adjustment for inflation). Disney CEO Michael Eisner's total compensation package went from $40 million to $576 million. By 1999, it was estimated that Berkshire Hathaway CEO Warren Buffet, Microsoft CEO Gates, and Microsoft retiree Paul Allen had a combined net worth of $156 billion—more than the combined gross national products of the world's 43 poorest nations. Thus the 1990s became known as the decade of greed and excess.

It is estimated that U.S. companies spent more than $1.2 trillion on information technology upgrades and software, hardware, and personnel additions from 1995 to 2000. The glowing predictions of the early 1990s led to heavy investment in the future of the Internet. Some researchers have even labeled the result of the Internet hype as "mass hysteria" because it resulted in a run-up of stock prices that defied logical explanation. From January 1994 to February 2000, the NASDAQ composite index—heavily oriented to the technology sector—rose from 776.80 to 4,696.69, a 605 percent increase. The Dow Jones Industrial Average—an average of 30 large companies on the New York Stock Exchange—rose about 10 percent a year from 1987 to 1995, then rose 15 percent a year from 1995 to 2000. The NASDAQ peaked March 10, 2000, at 5,048.62. The Dow's peak came on January 14, 2000, at 11,723.00.

The late 1990s had been a partial echo of the times just prior to the 1929 stock market crash. In both the '20s and the '90s many stocks were significantly overvalued, investors were buying a lot of stocks on margin, and a new technology was seriously influencing the economy in new ways.

The bear market began in 2000. When corporate profits in the late 1990s could not keep up with the rate of corporate spending, company valuations fell, stock prices fell, and what was referred to as the "Internet bubble" burst, sending markets significantly lower. Between January 2000 and July of 2002, the Dow dropped to the same level it would have reached had it followed the 10 percent growth rate of 1987 to 1995. Economists could label it as a "market correction" for the hype of the late 1990s, and despite the fact that a lot of people lost a lot of money, experts did not consider it to be a market crash.

Robert Shapiro, an undersecretary of commerce during the Clinton administration, put a positive spin on it, citing productivity studies that showed continuing gains after the tech bubble burst. "More than three-fourths of the increased productivity can be traced to the impact of these innovations and the high rates of investment in them," he wrote in a 2002 article for *Slate* headlined "What Is the Mother of Invention?" His point of view is that in the end the 1990s Internet hype did more good than harm.

ZEROING IN ON THE INTERNET AND SOCIETY IN THE EARLY 1990s
When stakeholders and skeptics were trying to make their points about the issues surrounding the Internet during the span of 1990 to 1995, they regularly utilized touchstone issues and personalities found in current events and popular culture to get their messages across.

Their news and pop-culture references ranged from the horrific bombing of the Alfred P. Murrah Federal Building in Oklahoma City, to O.J. Simpson's trial, to popular television shows, to cartoon characters, to major world figures of the time. For instance, Internet critic Clifford Stoll used the inanimate stars of the children's television series *Sesame Street* to illustrate his message. "Computers complement television," he wrote in his 1995 book *Silicon Snake Oil.* "No technological pathway—neither Muppets nor modem—leads directly to a good education."

In a 1993 interview in *Wired* magazine titled "Shock Wave (Anti) Warrior," futurist Alvin Toffler said new technology would bring "niche warfare," and he predicted, "That's exactly what we say we didn't do vis-a-vis Saddam (during the first Persian Gulf War in 1991), but what we will do. In fact there is a kind of dialectic here. . . . Many of the changes that we identify as carrying us into a third-wave civilization, or whatever, actually re-create pre-industrial conditions on a high-technology basis. And what you then see is individual assassination." Interestingly enough, in March 2003, after other attempts to remove Hussein failed, the United States launched an all-out war in Iraq to topple his regime. It was definitely not the niche warfare Toffler predicts for the future. Hussein was captured by U.S. troops months later, scruffy and alone in a hole in the ground, and taken to trial.

In a 1993 article titled "A Plain Text on Crypto Policy" in *Communications of the ACM,* Internet activist John Perry Barlow criticized the U.S. government for its stated fears regarding the ways criminals and terrorists might use the Internet, saying, "It seems to me that America's greatest health risks derive from the drugs that are legal, a position the statistics overwhelmingly support. And then there's terrorism, to which we lost a total of two Americans in 1992, even with the World Trade Center bombing, only six in 1993. I honestly can't imagine an organized ring of child molesters, but I suppose one or two might be out there. And the last time we got into a shooting match with another nation, we beat them by a kill ratio of about 2,300 to 1. Even if these are real threats, is enhanced wiretap the best way to combat them?" While government agencies had gathered some intelligence to indicate the possibility of terror attacks, no one expected the September 11, 2001, attacks on the World Trade Center and the Pentagon and the deaths of thousands.

The fall of the Berlin Wall, Nintendo, Dan Rather, Beavis and Butt-head, pogs, MTV, 1960s protest marches, Dr. Seuss, Larry King, The Unabomber, Elvis, John Updike, Cindy Crawford, Howard Stern, Julia Child, Luke Perry,

Swine Flu, "Baywatch," the Beatles, Siskel and Ebert, Rob Lowe, Ross Perot, Cary Grant, Frank Zappa, Rush Limbaugh, Meg Ryan, the Branch Davidians, and more are present in the predictions that follow.

POP/NEWS REFERENCES IN PREDICTIONS FROM 1990 TO 1992

The human dilemma is as it has always been, and we solve nothing fundamental by cloaking ourselves in technological glory. Even the humblest cartoon character knows this, and I shall close by quoting the wise old possum named Pogo, created by the cartoonist, Walt Kelley. I commend his words to all the technological utopians and messiahs present. "We have met the enemy," Pogo said, "and he is us."—Media critic **Neil Postman,** in a 1990 speech to the German Informatics Society in Stuttgart

A knowbot might respond to an instruction like "Find and display a good night-time picture of the Eiffel Tower" or "Play back the five most popular songs recorded by the Beatles." The knowbot would "travel" over the network, enter several computers it knows to contain this information, search around each using its syntax and conventions, combine the gleanings from these data sources into a single response and translate it into a format understood by the user's computer.—MIT's **Michael Dertouzos,** in a 1991 article for *Technology Review* titled "Building the Information Marketplace"

If the use of virtual communities turns out to answer a deep and compelling need in people, and not just snag onto a human foible like pinball or Pac-man; today's small online enclaves may grow into much larger networks over the next 20 years.—Writer/editor **Howard Rheingold,** in a 1992 essay on the EFF site titled "A Slice of Life in My Virtual Community"

Technology is transforming the face of democracy. . . . With modern technology, vastly more is possible. We are moving to a full-blown network of two-way television, in which anyone can be an orator or inquisitor. The network offers the pamphleteer and soap-box proselytizer not just a place in Speaker's Corner but the whole of Hyde Park. It gives the man on the couch precisely the same chance as Dan Rather to raise his hand and say something rude when the president steps into the room.—**Peter Huber,** a senior fellow at the Manhattan Institute, in his 1992 *Forbes* article "Telephone Democracy"

FROM FIELD OF DREAMS TO THE SATANIC VERSES IN 1993

It is inconceivable to the ABA that [a government-run, bank-deposit-tracking database] could only be used by the FDIC in deposit insurance-coverage functions. Such a database . . . would provide a wealth of information. . . . Like the baseball diamond in "Field of Dreams," build this database and they will come. Eventually, whether legally or illegally, they will gain access to this database. —Unnamed American Bankers Association spokesman, quoted in a 1993 *Wired* magazine article headlined "Big Brother Wants to Look in Your Bank Account"

A wired Armed Forces will be composed entirely of veterans—highly trained veterans of military cyberspace. An army of high-tech masters who may never have fired a real shot in real anger, but have nevertheless rampaged across entire virtual continents, crushing all resistance with fluid teamwork and utterly focused, karate-like strikes. This is the concept of virtual reality as a strategic asset. It's the reasoning behind SIMNET, the "Mother of All Computer Games." It's modern Nintendo training for modern Nintendo war.—Writer **Bruce Sterling,** in his 1993 *Wired* article "War Is Virtual Hell"

Blending our actual telephones and televisions will surely require a lot more than FCC approval. The video side of the information superhighway is a one-way street and will remain so for a long time—even if interactive boxes do someday let us send a few bytes worth of nay votes up the line toward Geraldo. —Science writer **James Gleick,** quoted in a 1993 *Wired* article titled "We Are the Wired: Some Views on the Fiberoptic Ties That Bind"

You'll be able to see "Terminator 2" at any time of the day or night. . . . I don't know that the world is going to be greatly changed [for the better by this opportunity].—Stanford communications professor **Henry Breitrose,** quoted in a 1993 *Wired* magazine article titled "Data Highways: Can We Get There from Here?"

By decade's end, we will look back at 1992 and wonder how a video of police beating a citizen [Rodney King] could move Los Angeles to riot. The age of camcorder innocence will evaporate as teenage morphers routinely manipulate the most prosaic of images into vivid, convincing fictions. We will no longer trust our eyes when observing video-mediated reality. Text will emerge as a primary indicator of trustworthiness, and images will transit the Net as multimedia surrounded by a bodyguard of words, just as medieval scholars routinely added textual glosses in the margins of their tomes.—Futurist **Paul Saffo,** in his 1993 *Wired* article "Hot New Medium: Text"

A decade ago, Ronald Reagan said in a speech that he was looking forward to the fall of the Berlin Wall and the end of the "Evil Empire." Most people thought he was talking about 200 to 300 years from now. Today, we know that it took only a few years. . . . When it ended, it ended in one big bang. That's what I see happening in the arena of education.—Discovery Institute researcher **Lewis J. Perelman,** quoted in a 1993 *Christian Science Monitor* article headlined "Will Technology Alter Traditional Teaching?"

A small-business man with a modest investment and a camcorder should be able to access the network to sell his wares. Or, if he wants, start his own version of "America's Funniest Home Videos."—EFF co-founder **Mitchell Kapor,** quoted in a 1993 *U.S. News & World Report* article titled "The Digital Democrat"

The new information-appliance PCs will not only be user-friendly, they'll also come in lots of different shapes and sizes. Think of them as the gadgets on "Star Trek." On the Enterprise, information appliances were everywhere. Handheld models diagnosed sick aliens and damaged transporters. There were computers that crew members talked into, typed into and sent messages back and forth over.—Writers **Catherine Arnst, Richard Brandt, Paul Eng, and Peter Burrows,** in their 1993 *Business Week* article "The Information Appliance"

There is an incredible potential for international friction in the fact that the Net is neither centralized nor censored. What happens when some college prankster posts chapters from "The Satanic Verses" to soc.culture.iranian? Not only do we not yet have solutions for these kinds of problems, but we also have relatively few policy experts who recognize that these are problems. But whenever problems begin to arise routinely from human interaction, it's a safe bet that lawyers will soon appear.—Electronic Frontier Foundation chief counsel **Mike Godwin,** in his 1993 *Internet World* article "The Law of the Net: Problems and Prospects"

SADDAM, "THE PRISONER," AND MILTON BERLE TOPICS IN 1994

FBI people . . . your idea of Digital Telephony is a scarcely mitigated disaster . . . you're going to be filling out your paperwork in quintuplicate to get a tap, just like you always do. . . . In the meantime, you will have armed the enemies of the United States around the world with a terrible weapon . . . raw and tyrannical Digital Telephony. You're gonna be using it to round up wise guys in street gangs, and people like Saddam Hussein are gonna be using it to round up democratic activists. . . .You're going to strengthen the hand of despotism around the world, and then you're going to have to deal with the hordes of state-supported truck bombers these rogue governments are sending our way after annihilating their own internal opposition by using your tools.—Writer **Bruce Sterling,** in his 1994 *Wired* magazine article "So, People, We Have a Fight on Our Hands"

While we're wiring people up we're destroying old cultures. It's great to have the idea of a global village, but it can resemble (the 1950s TV show) "The Prisoner," where everyone lives in a perfect village, but everyone is a number. Everyone is a prisoner.—Stanislaus State University professor **Tom Gentry,** quoted in a 1994 *Sacramento Bee* article headlined "Info-Culture Technology: Savior or Destroyer of Society?"

In January, Vice President [Al] Gore had promised that the White House would work to ensure that the NII [National Information Infrastructure] would "help law enforcement agencies thwart criminals and terrorists who might use advanced telecommunications to commit crimes." . . . His pledge went unnoticed by the mainstream press. Notwithstanding that it fell on reporters' deaf ears, Gore dropped a bombshell. Forget Ross Perot's NAFTA-inspired "giant sucking sound." This was the dull "thump" of Law Enforcement running over the privacy rights of the American public on its way—at the on-ramp??—to the information superhighway. The real crime is that the collision barely dented the damn fender.—Technology writer **Brock Meeks,** in his 1994 *Wired* article headlined "If Privacy Isn't Already the First Roadkill Along the Information Highway, Then It's About to Be"

My software surrogates can . . . serve as my semiautonomous agents by tirelessly performing standard tasks. . . . A more maliciously conceived one might be programmed to roam the digital highways and byways, looking for trouble—for opportunities to corrupt the files of my enemies, to plunder valuable information, to eliminate rival agents, or to replicate itself endlessly and choke the system. Fritz Lang got it wrong: the robots in our future are not metallic Madonnas clanking around "Metropolis," but soft cyborgs slinking silently through the Net. The neuromans of William Gibson are a lot closer to the mark.—MIT professor **William Mitchell,** in his 1994 book *City of Bits*

An industry will grow up around individuals licensing their points of view for use in context engines in exchange for usage royalties. Imagine being able to give your news agent the personality and perspective of Walter Cronkite, Howard Stern or John Updike, or consult the software-doubles of Siskel and Ebert for advice on cool movies to view. Just as talk show hosts have become the movers and shakers of post-network TV today, individuals with unique points of view could become the superstars of cyberspace, their personalities immortalized in software traversing the Web.—**Paul Saffo,** in his 1994 *Wired* article "It's the Context, Stupid"

I don't believe "Beavis and Butt-head" on demand is going to drive (sales on) the information highway. Every child I know already has an Encyclopaedia Brittanica within reach as they watch MTV. So why is wiring them up to the Library of Congress going to change things?—Sun Microsystems CEO **Scott McNealy,** quoted in a 1994 *San Francisco Chronicle* article headlined "Business May Drive Information Highway"

Marketers should get back to their TV roots. . . . The advertisers who will get rich by brilliantly exploiting online services won't do it by feeding us specs, brochures, and blatantly bad propaganda, but by being sponsors. Let artists create content. Smart advertisers will make it big by sponsoring the digital equivalent of the "The Milton Berle Show" (aka "The Texaco Star Theater"). . . . Online advertising needs a 21st-century Uncle Miltie who can captivate the wired masses with amusing shtick while shamelessly hyping the sponsor who paid for it all. Adam Curry, meet Jack Benny.—Public relations consultant **Chris Clark,** in his 1994 *Wired* article "And Now, a Word from Our Sponsors"

In a truly interactive system, product pitches could be recorded in advance and stored, much like voice-mail messages today, in powerful central comput-ers called servers. Then anyone interested in, say, rodeo belt buckles and cubic-zirconia adornments could zip straight to the relevant video. Interactive video is a medium made for merchandising. Want to see how those Eddie Bauer waders look on Cindy Crawford? Just click. In the market for a vintage El Camino like Bill Clinton owned in the 1970s? Click and ye shall find.—Writer **Stratford Sherman,** in his 1994 *Fortune* magazine article "Will the Information Superhighway Be the Death of Retailing?"

O.J., MCDONALD'S, LEVIS, AND THE UNABOMBER TOPICS IN 1995

Unlike the craze for baggy jeans and Pogs, this isn't just some passing fad. Market researchers predict the number of kids online could triple by the end of 1998. So when the big bills come—and they will—parents will have only themselves to blame for bringing computers and online services home with them.—Technology writer **Connie Guglielmo,** in her 1995 *PC Week* article "Just Child's Play"

The history of U.S. technology is the history of a recurring U.S. dream: new inventions will empower the individual more than the corporation. . . . A current variant is the prediction that the anarchic Internet will turn people into media makers and kill off more restrictive commercial services like America Online and CompuServe. This prediction has two flaws. First, in a society satisfied by "Baywatch" reruns, few people will produce or consume the amateur media. Second, no matter how nice the Web viewer, the unstructured Internet will always be much harder to use than an online service.—Writer **Steve Steinberg,** in his 1995 *Wired* magazine article "Hypelist: Death of Online Services"

The Oklahoma bombing has fueled government hysteria about the evils of free speech, but that's not the half of it. With the FBI "reinterpreting" current regulations to investigate those it considers "potential terrorists," we'd do well to remember our history. When we approve broad powers for organizations with a proven track record of ignoring civil liberties, we do so at the expense of all our liberties.—*Wired* magazine editors, 1995, in a column titled "Flux: Things Are Heating Up"

The amount of time I save by being able to find a piece of information on-line is almost exactly negated by how much time I waste every day by being on-line. . . . The thing that I do want a machine to solve is something that a machine can't solve namely, "give me only the e-mail that is essential in my work."

For me to program that is for me to know what is essential in my work, and I don't know that. It changes every 20 minutes. As an example, I would never in my life program an artificial intelligent agent to tell me what's happening in Oklahoma City. Yet, all of a sudden last April, a bomb goes off and everything in Oklahoma City is important and essential.—Physicist and Internet critic **Clifford Stoll**, quoted in a 1995 *Computerworld* article titled "An Interview with Cliff Stoll"

Digital media could make it possible for people to interact—maybe even changing each other's minds in the process—something traditional media inhibit through their addiction to objectivity, spokespeople, and sensationalism. . . . Online news suggests a forum in which it would be easier for fragmented political or racial groups to begin . . . teaching members of [many] tribes how to communicate and providing them with the simple means of doing so. . . . If one tenet of our age is that information wants to be free, its companion is that media want to tell the truth. Neither information nor media get what they want much of the time; this is one of the great ironies of the information revolution and the sad legacy of the O.J. Simpson trial.—Media critic **Jon Katz**, in his 1995 *Wired* article titled "Guilty"

Cyberspace, not mainstream media, would be [Thomas] Paine's home now. Commentary has virtually vanished from TV, and the liveliest newspaper Op-Ed pages are tepid compared to Paine's tirades. But online, millions of messages centering on the country's civic discourse are posted daily, in forums teeming with the kind of vigorous democratic debate and discussion that Paine and his fellow pamphleteers had in mind. Gun owners talk to gun haters, people in favor of abortion message people who think abortion is murder, journalists have to explain their stories to readers, and prosecution and defense strategies in the O.J. Simpson trial are thrashed out.—Media critic **Jon Katz**, in a 1995 essay for *Wired* magazine titled "The Age of Paine: Thomas Paine . . . Should Be Resurrected as the Moral Father of the Internet"

It's depressing that there's a McDonald's at Singapore airport and a Gap on every street in the world. . . . Diversity is valuable if something goes wrong. [With global communications like the Internet,] we're taking down all the barriers, and I'm not sure that's a good thing.—Best-selling author **Michael Crichton**, quoted in a 1995 *Seattle Times* article headlined "'Lost World'

Suggests a Bad Attitude Killed the Dinosaurs and Humans May Be Headed Down the Same Road"

Just as information technology now allows Levi Strauss & Co. to offer jeans that are both mass-produced and custom fitted, information technology will bring mass customization to learning. Multimedia documents and easy-to-use authoring tools will enable teachers to "mass-customize" a curriculum. As with blue jeans, the mass customization of learning will be possible because computers will fine-tune the produce—educational material, in this case—to allow students to follow somewhat divergent paths and learn at their own rates.—**Bill Gates,** in his 1995 book *The Road Ahead*

Some Internet users say putting the manifesto on the World Wide Web may help the FBI find the Unabomber. "This is something a lot of us have been pushing for a couple of months," said Stewart Brand. . . . Brand says the virtual community . . . could ferret out the identity of the Unabomber. Brand says it reminds him of a Frank Zappa concert. Someone threw a bottle of beer at Zappa, and the musician stopped the concert until the culprit was found out. As Brand describes it, the attention of the crowd began in the far reaches of the auditorium. They looked toward the origin of the thrown bottle. Then the next wave of people looked toward the spot. Then the people around the bottle thrower looked to the spot. Finally only one person wasn't looking at anyone else. The security guards hustled him out. "With any luck," Brand says, "this could happen on the Net."—Reporter **Mark Fisher** quotes WELL co-founder Stewart Brand in a 1995 *Chicago Sun-Times* article titled "Unabomber's Words Find an Audience"

Computing corduroy, memory muslin, and solar silk might be the literal fabric of tomorrow's digital dress. Instead of carrying your laptop, wear it. . . . The important point is to recognize that the future of digital devices can include some very different shapes and sized from those that might naturally leap to mind. . . . Computer retailing of equipment and supplies may not be limited to Radio Shack and Staples, but include the likes of Saks and stores that sell produces from Nike, Levi's, and Banana Republic. In the further future, computer displays may be sold by the gallon and painted on, CD-ROMs may be edible, and parallel processors may be applied like suntan lotion. Alternately, we might be living in our computers.—**Nicholas Negroponte,** in his 1995 book *Being Digital*

Where is the place for politics in this brave new world, when leaving the Net becomes as unthinkable as giving up breathing? . . . As the warp and the woof draw ever tighter, the feelings of claustrophobia and manipulation that result may indeed trigger a new politics in the midst of digital culture: the networked equivalent of the Branch Davidians, where the ultimate political gesture is one of withdrawal and self-marginalization.—Writer **Jay Kinney**, in his 1995 *Wired* magazine essay titled "Is There a New Politics Emerging in the Net/Cyberspace/Digital Culture?"

Your phone or computer will be able to generate a lifelike digital image of your face, showing you listening or even talking. You really will be talking—it's just that you've taken the call at home and are dripping wet from the shower. As you talk, your phone will synthesize an image of you in your most businesslike suit. Your facial expressions will match your words (remember, small computers are going to get very powerful). . . . If you are talking to someone you've never met, and you don't want to show a mole or a flabby chin, your caller won't be able to tell if you really look so much like Cary Grant (or Meg Ryan) or whether you're getting a little help from your computer.—Bill **Gates**, in his 1995 book *The Road Ahead*

What the hell do you think the Internet is for? It isn't a replacement for radio, TV, and telephones. It's for exchanging information, not free phone calls. . . . From a store-and-forward perspective, the Internet was a new chapter in the history of human intelligence. It was supposed to lead us somewhere higher, to something better. That some company could then turn this spiritual adventure into another vehicle to support eighth-grade schoolgirls babbling about Luke Perry was nothing short of criminal. Yet all this grousing delayed the spread of the technology only a few hours. . . . the floodgates were open to the chattering hordes.—Technology writer **Fred Hapgood**, in his 1995 *Wired* magazine article "IPhone: Will Telephony on the Net Bring the Telcos to Their Knees?"

The persona of a machine makes it fun, relaxing, usable, friendly, and less "mechanical" in spirit. . . . You will be able to purchase personality modules that include behavior and style of living of fictitious characters. You will be able to buy a Larry King personality for your newspaper interface. Kids might wish to surf the Net with Dr. Seuss. . . . We will see systems with humor, systems that nudge and prod, even ones that are as stern and disciplinarian as a Bavarian nanny.—**Nicholas Negroponte,** in his 1995 book *Being Digital*

In the MIT Media Lab's version of the future, people will customize their computer news "guide" once, and then the day-to-day work will be done automatically. This robot will go out and get the news—not the news that a professional journalist would choose, but the specific kinds of topics that the consumer says she wants. Journalists, if they're smart, will offer continual information guidance that obviates the need for such robots. To do this, they may not have to be as entertaining or as ideological as Rush [Limbaugh's] reports, but they will have to be more accurate, more relevant, and more attuned to their audiences than most are today.—Communications researcher **Ellen Hume,** in her 1995 paper titled "Tabloids, Radio, and the Future of News"

Cash is a dubious thing. . . . Cyberspace is where the bank keeps your money and to a real extent it's where the stock market happens. I'm waiting for the Three Mile Island of computer banking—some unspeakable meltdown, although I'm hoping it really doesn't come along. That will be when we discover the extent of our reliance on computation.—Sci-fi author **William Gibson,** quoted in a 1995 *New York Times* article titled "Online with William Gibson; Present at the Creation, Startled at the Reality"

In a couple of years everyone will be able to put their head on my body. They'll be able to decide, "I choose to ask Ed Asner this question instead of the one Conan asked." No one will ever see me doing my talk show. They'll see themselves doing their own talk show. They'll be able to choose Rob Lowe as a guest instead of Ed Asner. They'll get a computer-generated Rob Lowe. I'll become a guy who just installs modem jacks.—NBC-TV late-night talk show host **Conan O'Brien,** quoted in a 1995 *New York Times* article headlined "The Talk of Cyberspace"

The same technology that will put every "Star Trek" episode ever made within reach of a few clicks of your set-top device will also give you access to educational resources beyond those available on even the finest university campus today. Imagine a virtual university where the best lecturers in the world are on tap at the moment you are ready to concentrate on learning.—**Jean Jipguep,** chairman of the International Telecommunication Union

Board, in his keynote address to the 1995 Internet Society International Networking Conference

In the future, companies will give you stuff. For instance, Domino's might give you a little machine with two buttons: pizza with cheese, pizza with pepperoni. You throw it on top of your refrigerator and you come home late at night. You're so lazy you won't even make a phone call—you press one button! There's a cellular modem connected by cellular digital packet data, a new standard for sending data over cell lines to the cellular data network, back to Domino's. Fifteen minutes later, there's a pizza.—Forrester Research president **George Colony,** quoted in a 1995 *Wired* magazine article titled "'Golden Guts' Colony Delivers Some Really Educated Guesses"

7

Nothing Is Certain but Death and Taxes

(And some predictions—including the death of taxes—may have been premature, while many "deaths" may come to pass)

People expect the Internet to transform our world in myriad ways, and one of the classic manners in which people have always expressed the likelihood of change is by predicting the death of existing tools, conventions, or social structures.

Researchers who have studied the diffusion of innovations (led by Everett Rogers, whose definitive book on the topic has been updated a number of times since it was first released in 1962) say that users of a new tool are naturally bound to pass judgment on that tool and then share their opinion. Users and other stakeholders—experienced with the tool or not—will identify what they foresee to be the individual and social consequences of the tool.

The first people to express their opinions in the diffusion of a new tool have been classified as innovators, change agents, reactionaries, iconoclasts, or early adopters. *Innovators* are the inventors of the tool—in the case of the Internet, this would be the many pioneers who built it, including Vinton Cerf and Tim Berners-Lee. *Change agents* are idea brokers for the innovation; they promote mostly the positive aspects of the change to come thanks to the innovation.

Entrepreneurs, researchers, and people in government were change agents for the Internet. *Reactionaries* resist the adoption of the innovation, preferring the status quo prior to the innovation. Included in this group during the early days of the Internet in the 1990s would be opponents such as Sven Birkerts and Kirkpatrick Sale. *Iconoclasts* are silent partners to the innovator, hoping for change for the better—they are often journalists or social gadflies, as in the case of the Internet, such as Howard Rheingold and Bruce Sterling. *Early adopters* are also called *transformers.* They become users of the new tool out of excitement and hope for a positive change. The myriad "plain folks" who were the first to explore virtual communities online would fit into this category.

Between 1990 and 1995, the innovation of the Internet brought an unprecedented outpouring of opinions from stakeholders and skeptics representing all of the groups listed above. These opinion leaders, of large or small profile, were instrumental in diffusing the innovation known as the Internet. Since then, the Internet has become the most effective worldwide super-diffusion tool, allowing anyone to share information about ensuing innovations to a worldwide audience at no cost.

As the initial wave of awe regarding the potential of the Internet began to hit home, people happily, fearfully and/or warily predicted the death of taxes, books, the CD, the recording industry, TV, e-mail, mainframe computers, copyright and patent law, big corporations, political parties, conventional schools, commuting to work, major urban centers, and all institutions, behaviors, and values that had developed since the 18th century. They predicted a paperless society and the extinction of the human race after a takeover engineered by intelligent machines; did you know you could be a museum piece yourself?

The following statements, culled from thousands of prognostications made between 1990 and 1995, include predictions of the beginning of the end, for good or for bad, depending upon your point of view.

SLATED FOR EXTINCTION:

Phone Numbers Scrawled in Public Bathrooms

"For a good time, get online" may soon replace the more familiar graffiti scribbled in bathroom stalls around the world.—Writer **Jay Dougherty**, in a 1994 *Rocky Mountain News* article titled "Sexy Conversations Luring Many Users to Online Services"

Industrial Civilization

If the edifice of industrial civilization does not eventually crumble as a result of a determined resistance within its very walls, it seems certain to crumble of its own accumulated excesses and instabilities within not more than a few decades, perhaps sooner, after which there may be space for alternative societies to arise.—**William Mitchell,** in his 1994 book *City of Bits*

Political Parties

The fundamental thing [the Net does] is to overcome the advantages of economies of scale . . . so the big guys don't rule. . . . [Organized political parties won't be needed if open networks] enable people to organize ad hoc, rather than get stuck in some rigid group.—Consultant **Esther Dyson,** quoted in a 1994 article for *Wired* magazine headlined "The Merry Pranksters Go to Washington"

Tax Collection

It is imaginable that, with the widespread use of digital cash and encrypted monetary exchange on the Global Net, economies the size of America's could appear as nothing but oceans of alphabet soup. Money laundering would no longer be necessary. The payment of taxes might become more or less voluntary.—EFF co-founder **John Perry Barlow,** in a 1993 article for *Communications of the ACM* titled "A Plain Text on Crypto Policy"

One thing for sure, long-term, [online anonymity] nukes tax collection. Without a doubt, this stuff is unbreakable. Encryption always wins.—Self-described "crypoanarchist" **Tim May,** quoted in Kevin Kelley's 1994 book *Out of Control: The New Biology of Machines, Socials Systems, and the Economic World*

If total public cryptography and lots of financial transactions come to the Net, will you pay taxes in the future? You won't. This is one terrifying fantasy from the government standpoint. . . . A whole lot of financial activity basically goes black, goes underground. And then you can't tax transactions, you can't track transactions. All you've got left to tax basically is possessions at that point and so you may see . . . property taxes going up and sales taxes disappearing.—Internet pioneer **Stewart Brand,** in a 1995 PBS-TV interview on a showed titled *High Stakes in Cyberspace*

Restriction of Copying

Let the copies breed. Whatever it is that we are constructing by connecting everything to everything, we know the big thing will copy effortlessly. The I-way is a gigantic copy machine. It is a law of the digital realm: anything digital will be copied, and anything copied once will fill the universe. Further, every effort to restrict copying is doomed to failure.—*Wired* editor **Kevin Kelly,** in a 1994 article for the *London Guardian* headlined "In 2004 We'll All Live on the Internet with Silicon Valley Visionaries"

Copyright and Patent

What was previously considered a common human resource, distributed among the minds and libraries of the world, as well as the phenomena of nature herself, is now being fenced and deeded. It is as though a new class of enterprise had arisen that claimed to own the air. . . . dancing on the grave of copyright and patent will solve little, especially when so few are willing to admit that the occupant of this grave is even deceased, and so many are trying to uphold by force what can no longer be upheld by popular consent.—EFF co-founder **John Perry Barlow,** in a 1994 *Wired* magazine article titled "The Economy of Ideas"

We are clueless about the ownership of bits. Copyright law will disintegrate. . . . Bits are bits indeed. But what they cost, who owns them, and how we interact with them are all up for grabs.—MIT Media Lab co-founder **Nicholas Negroponte,** in a 1995 *Wired* magazine column headlined "Being Digital: A Book (P)review"

The Telephone and Television Industries

Revenues from telephones and televisions are currently at an all-time peak. But the industries organized around these two machines will not survive the century. . . . All the assumptions of telephony will have to give way to radically different assumptions. Telephony will die. . . . TV ignores the reality that people are not inherently couch potatoes; given a chance, they talk back and interact. People have little in common except their prurient interests and morbid fears and anxieties. Necessarily aiming its fare at this lowest-common-denominator target, television gets worse and worse every year. Television is a tool of tyrants. Its overthrow will be a major force for freedom and individuality, culture and morality. That overthrow is at hand.—Technology consultant **George Gilder,** in his 1994 book *Life After Television*

[Talking about a convergence of television and computers into one Internet device is] a bit like talking about the convergence of the horseless carriage and the modern automobile. The proper word is "replacement." The computer is taking over the television.—Intel CEO **Andy Grove,** quoted in a 1994 *London Observer* article headlined "Superhighway or Dead End?"

The Mass Media as We Know Them

To my mind, it is likely that what we now understand as the mass media will be gone within 10 years. Vanished, without a trace. . . . Who will push *The New York Times?* The answer, I think, is technology. . . . Consumers will naturally want better information. They'll demand it, and they'll be willing to pay for it. There is going to be—I would argue there already is—a market for extremely high-quality information.—Author **Michael Crichton,** in a 1993 essay titled "Mediasaurus: Today's Mass Media Is Tomorrow's Fossil Fuel"

The Bureaucratic Organization

Turning the economics of mass production inside out, new information technologies are driving the financial costs of diversity—both product and personal—down toward zero, "demassifying" our institutions and our culture. Accelerating demassification creates the potential for vastly increased human freedom. It spells the death of the central institutional paradigm of modern life, the bureaucratic organization.—**Esther Dyson, George Gilder, Jay Keyworth, and Alvin Toffler,** in their 1994 *Magna Carta for the Knowledge Age*

E-mail

In the next decade, electronic mail is dead.—Futurist **Paul Saffo,** quoted in a 1994 *New York Times* article headlined "The Rise and Swift Fall of Cyber Literacy"

CDs, Faxes, Snail Mail

With more bandwidth, do we need CDs at all? . . . The worst thing about the arrival of magazines and reports is the ability to find the magazine and article once you've read it. Did you throw it away? Where is it filed? What issue is it in? I'd gladly pay extra for online access to a magazine so that I don't have to file it. In some cases, I'd pay extra to never get it at all and simply access it when I need to. Bills is my favorite one to come electronically. . . . Merchants send us their bills via e-mail. And thank god, faxes are disappearing. We owe it to the world to get rid of faxes. And, finally, personal letters. For many of us letters have almost disappeared. We don't owe it to the world to get rid of personal letters because they're nice things. In a lot of cases, Internet brings people closer together because they're on already and using it for personal communication is easy and natural.—Pioneering computer scientist **Gordon Bell,** in his InternetWorld 1995 keynote speech, titled "It's Bandwidth and Symmetry, Stupid!"

Institutions, Behaviors, Values of the 18th–20th Centuries

The transforming power of the Internet, and all of its possible successor netwoven communication technologies, are in the process of completely destroying all of the institutions, behaviors, and values which arose around the industrial technologies of the 18th, 19th, and 20th centuries—economic, military, political, cultural—just as industrial technologies destroyed—or at least marginalized and substantially changed—the institutions, behaviors, and values of pre-industrial, agricultural societies.—Futurist **Jim Dator,** in a 1995 speech for the International Conference on Development, Ethics, and the Environment titled "Coming Ready or Not: The World We Are Leaving Future Generations"

As the electronic revolution merges with the biological evolution, we will have—if we don't have it already—artificial intelligence, and artificial life, and will be struggling even more than now with issues.... During the 21st century all historically experienced human processes—agriculture, industry, commerce, education, you name it—will come to an end.... What is actually happening . . . is the merger of four information societies into one: the 4-billion-year-old genetic information society; the 10,000-year-old cultural information society; the 3,000-year-old civilizational society; and the 250-year-old industrial information society, all merging in the 21st century into one new "coming information society."—Futurist **Jim Dator,** in a 1993 keynote speech at the WFSF World Conference titled "Dogs Don't Bark at Parked Cars"

It is going to destroy vast layers of our economy and make available a presence in the marketplace for very small companies, one that is equal to very large companies.—Apple and NeXT CEO **Steve Jobs,** in a 1995 video interview for the Smithsonian Institution Oral and Video Histories Series

Emoticons

If you want a sentence to end with a chuckle to show that its meaning is intended to be humorous, you might add a colon, a dash, and a parenthesis. This composite symbol, :-), if viewed sideways, makes a smiling face. . . . These "emoticons," which are half cousins of the exclamation point, probably won't survive the transition of e-mail into a medium that permits audio and video.—Microsoft CEO **Bill Gates,** in his 1995 book *The Road Ahead*

Cable Company Monopolies

More deregulation awaits us. This will mean the end of cable company monopolies, and the beginning of a new era of computer-integrated interactive entertainment, consumerism, and education.—Architect and technology visionary **Michael Benedikt,** in a lecture at the 1992 New Urbanism Symposium at Princeton University titled "Cityspace, Cyberspace and the Spatiology of Information"

Recording Companies

We could do without record companies entirely. Record companies . . . distribute pre-recorded copies of music. . . . But listeners making copies for themselves or their friends do not consume this service; they use only the work of the musicians and composers. . . . We can promote music more effectively by making any one musician's share of the tax revenues taper off as copies increase. For example, we could calculate an "adjusted number of copies" beyond which revenue increases more slowly than the actual number, following a prescribed mathematical function. The effect of tapering off will be to spread the money more widely, supporting more musicians at an adequate standard of living. This encourages diversity, as copyright was supposed to do.—**Richard Stallman,** president of the Free Software Foundation, head of the GNU project, and a MacArthur Foundation Fellow, in a 1993 *Wired* magazine article titled "Copywrong"

The NII may be the Holy Grail for lovers of pre-recorded music, but without change in the copyright law it would also be a death knell to our industry. . . . A transmitter would simply be able to procure one copy of a copyrighted work, and then transmit it to thousands of users at the touch of a button. One can hardly imagine a more bleak future for our company.—**Lawrence Kenswil,** a vice president with MCA Music Entertainment Group, in a 1993 article in *Billboard* titled "Trade Wants C'Right Assurances as Info Highway Is Paved"

In the future, CD's may be abolished altogether; consumers will be able to listen to whatever song or album they want by ordering it (for a fee) on their cable television box. Record labels would no longer be necessary: anyone with a song and a computer could just put it online.—Writer **Neil Strauss,** in a 1995 *New York Times* article headlined "Records of the Future: At Your Fingertips"

Ads on Radio Stations

I'm listening to the FM radio and I hear a song I like. I press a question mark and I instantly see the name of the artist and the other songs on the album and the price of the album. And there's the damn BUY button, and the next day a CD will show up in my house. . . . I can buy $7 CDs instead of $15 CDs. And you know what's cool for the radio station? The radio station gets tagged, because this device knew what station you were listening to, and what time. So in your order, they include the information, and the CD guys can send a fee over to the radio station to subsidize them for playing it. It could lead to radio stations with no commercials.—Hypercard inventor **Bill Atkinson,** in a 1994 *Wired* article titled "Bill and Andy's Excellent Adventure II"

Checks, Cash, Coins, and ATMs

About 60 billion checks are written annually in the United States and, according to check printer Deluxe Corporation, the numbers will decline in the next century. Company officials say they are in the process of closing plants and cutting staff to begin focusing on other ventures.—Writer **Rose Aguilar,** in a 1995 CNET News.com article titled "Check Printers Fear Online Banking"

The now-ubiquitous ATMs (in their role as cash dispensers, at least) will become obsolete if coins and bills are eventually eliminated. This is a fairly straightforward technical possibility; a combination of network transfers, checks, credit cards, debit cards, ubiquitous point-of-sale terminals, and replacement of coin-operated gizmos like parking meters with electronic card-reading devices clearly could yield a cash-free society.—**William Mitchell,** in his 1994 book *City of Bits*

VCRs, Videocassettes, and the Video Rental

The first entertainment atoms to be displaced and become bits will be those of videocassettes in the rental business, where consumers have the added inconvenience of having to return the atoms and being fined if they are forgotten under a couch ($3 billion of the $12 billion of the U.S. video rental business is said to be late fines). Other media will become digitally driven by the combined forces of convenience, economic imperative, and deregulation. And it will happen fast.—**Nicholas Negroponte,** in his 1995 book *Being Digital*

Universities

There is nothing natural about taking 18-year-olds out of the world for three years into the cloistered halls of academia. . . . with new technologies relentlessly redefining the way we work and live, it may not merely be an anachronism to continue to embrace the model of the traditional residential university as the primary locus of learning—it may arguably be an impediment to appropriate learning and ultimately a threat to growth, both economic and personal. If structured high-quality learning materials are available online to whoever has access to a computer and modem, without constraints of time and place, then the traditional residential teaching university becomes—from the students' perspective at least—largely redundant. —Kingston University information systems specialist **Chris Hutchinson,** in his 1995 *Journal of Computer-Mediated Communication* article "The 'ICP Online'"

Book Stores and Books

If someday in the future anybody can get an electronic copy of any book from a library free of charge, why should anyone ever set foot in a bookstore again?—Writer/editor **John Browning,** in a 1993 *Wired* article titled "Libraries Without Walls for Books Without Pages"

The novel . . . as we know it, has come to its end. . . . True freedom from the tyranny of the line is perceived as only really possible now at last with the advent of hypertext, written and read on the computer, where the line in fact does not exist unless one invents and implants it in the text. . . . With its webs of linked lexias, its networks of alternate routes . . . hypertext presents a radically divergent technology. . . . Hypertext reader and writer are said to become co-learners and co-writers, as it were, fellow-travelers in the mapping and remapping of textual (and visual, kinetic and aural) components, not all of which are provided by what used to be called the author.—Online literature pioneer **Robert Coover,** in a 1994 *New York Times* article titled "The End of Books"

This is not just another format. In the long term, digital media will fundamentally destabilize the way [encyclopedia publishers] do business. Usually, people talk about the revolution in digital media in terms of putting interactivity within the product itself. But the real revolution is in the market.—Joe Esposito, president of Encyclopaedia Brittanica, quoted in a 1995 *Wired* article titled "Encyclopaedia Brittanica Online?"

In the past, education adapted the mind to a very restricted set of available media; in the future, it will adapt media to serve the needs and tastes of each individual mind. . . . The Knowledge Machine (a metaphor for much more varied forms of media) will provide easier access to richer and fuller bodies of knowledge than can be offered by any printed encyclopedia.—Artificial-intelligence researcher **Seymour Papert,** in his 1993 *Wired* article titled "Obsolete Skill Set: The 3 R's—Literacy and Letteracy in the Media Ages"

Mainframes, Minicomputers, Servers, and Workstations

Individual low-cost, high-powered PCs, such as Compaq Computer Corp.'s ProLiant, combined with Windows NT, SQL-based databases and a single communications network will form the heart of the scalable computer. You can say good-bye to mainframes, proprietary minicomputers, servers and workstations.—**Gordon Bell,** in a 1995 *Computerworld* article headlined "The View from Here: Gordon Bell Previews a Future in Which Plugging in to a Worldwide Network Is as Easy as Getting a Dial Tone"

The "Killer Application"

The next thing for the Web is the death of the concept of the killer application. It will be killer content. The idea of an application will disappear over time. There is one possibility. . . . There will be a whole mingling of components of software which won't be grouped into lumps like applications. Even the operating system will become less significant. What you will be interested in in your operating system is something which will be small and fast and get out of the way quick.—World Wide Web inventor **Tim Berners-Lee,** in a 1995 *Computer Reseller News Industry* article titled "Web Inventor Berners-Lee Speaks Out on Internet Future"

AND NEAR-DEATH PROGNOSTICATIONS . . .

Business Travel

Teleconferencing—in which you can share data, images and speech—will make the concept of much business travel redundant. . . . Already airlines are scaling down their expectations of numbers of business travelers toward the end of the century, and it's the computer that is to blame.—**Andy Grove,** in a 1994 *London Observer* article titled "Superhighway or Dead End?"

Privacy

Privacy is under siege. The issues are especially vital today as more and more of our privacy is stripped away. And the notion that information can be kept secret to any degree may simply vanish in cyberspace.—Lawyers **Caroline Kennedy and Ellen Alderman,** in their 1995 book *The Right to Privacy*

Big Cities

Will the development of cyberspace precipitate a migration away from the crime-ridden big cities back to rural living, a trend which would greatly affect state and local planning? This is possible if people are able to send their children off each morning to a virtual school or university and then report to work in a virtual office where they interface with co-workers hundreds or even thousands of miles apart, then drop into a virtual shopping mall at lunchtime to handle their more elaborate shopping needs, get together with friends after work at a virtual cafe, and then download the news, book, television program or film of their choice to pass the evening hours.—Writer **Blake Harris,** in a 1995 *Government Technology* article titled "Cyberspace 2020"

Software will become friendlier, and companies will base the nervous systems of their organizations on networks that reach every employee and beyond, into the world of suppliers, consultants, and customers. The result will be companies that are more effective and, often, smaller. In the longer run, as the information highway makes physical proximity to urban services less important, businesses will decentralize and disperse their activities, and cities, like companies, may be downsized.—**Bill Gates,** in his 1995 book *The Road Ahead*

Great cities will hollow out, as the best and brightest in them retreat to rural redoubts and reach out to global markets and communities.—**George Gilder,** in a 1994 *National Review* article headlined "Net Gains: Information, Technology, and Culture"

Traditional Downtowns and Mom-and-Pop Stores

Cyberspace is going to finish what Wal-Mart started. Interactive shopping via computer networks is going to put more traditional downtowns and more mom-and-pop stores out of business.—Public-policy researcher **Richard Sclove,** quoted in a 1994 *New York Times* article headlined "Staking a Claim on the Virtual Frontier"

Traditional Corporations

We're going to see a widespread disintegration of U.S. business and the emergence of very different corporate entities.—Executive **John Hagel III,** quoted in a 1993 *Fortune* magazine article titled "Boom Time on the New Frontier"

The Existing Business Structure

Within the next few years, business-to-consumer electronic commerce could begin to replace much of the world's existing business infrastructure. —Sun Microsystems CEO **Scott McNealy,** quoted in a 1995 *San Francisco Chronicle* article headlined "New Sun Micro Products Link with the Net"

File Drawers, Storage Boxes, and More

When information appliances are connected to the highway, there will be less need for many physical things—reference books, stereo receivers, compact discs, fax machines, file drawers, and storage boxes for records and receipts. A lot of space-consuming clutter will collapse into digital information that can be recalled at will.—**Bill Gates,** in his 1995 book *The Road Ahead*

Paper-based "Paperwork"

Legal, governmental, medical, and . . . nearly every service that uses paper today could do business via the NII. . . . The process of moving around the paper forms and letters that constitute business mail, for example, consumes tens of billions of dollars annually.—MIT's **Michael Dertouzos,** in a 1991 *Technology Review* article headlined "Building the Information Marketplace"

Election Polling Places

Polls show that significant majorities favor the idea of national referendums—higher taxes—yes or no? Like church dogma, our civic religion now dictates that even when the people seem wrong, they are—by definition—right. Popul infallibility. Soon this hyperdemocratic impulse will be harnessed to irresistable technologies. . . . Interactive voting is closer than we know. With as many as 50 percent of American homes expected to have a modem within the next five years, the decline of the polling place may be at hand.—Political writer **Jonathan Alter,** in a 1995 *Newsweek* magazine essay headlined "The Couch Potato Vote: Soon You'll Be Able to Vote from Home—But Should You?"

Phone Companies

The Internet, now a boon for telephone companies, could well become their bane. . . . If I had stock in the telephone company. I would sell it. —Technology writer **Fred Hapgood,** in a 1995 *Wired* article titled "IPhone: Will Telephony on the Net Bring Telcos to Their Knees?"

The phone companies that survive will become cellular phone companies. "Anyone, Anywhere, Anytime" is a good motto for a 21st century phone company. There will be flat rates for so-called "long-distance." Any nation or PTT (postal, telephone, and telegraph company) which tries to cling to current long-distance telephony billing practices will see their economy destroyed by others with more enlightened policies.—Technology writer **Bruce Sterling,** in a 1995 e-mail interview with Telecommunications International that was posted on its website with the headline "Dropping Anchor in Cyberspace"

Court-ordered Wiretaps

As the information superhighway continues to expand into every area of society and commerce, court-ordered wiretaps and seizures of records could become tools of the past, and the information superhighway a safe haven for criminal and terrorist activity.—Internet security researcher **Dorothy Denning,** in a 1994 *Journal of Criminal Justice Education* article titled "Crime and Crypto on the Information Superhighway"

Buildings

Cyberspace can be seen as extending an inexorable process that began a long, long time ago and which gained new impetus earlier this century, namely the dematerialization of buildings.—**Michael Benedikt,** in a 1992 lecture at the New Urbanism Symposium at Princeton University titled "Cityspace, Cyberspace and the Spatiology of Information"

The Middle Stage of Childhood

When I was a child, no one challenged the three-stage model (requiring education in elementary, middle, and high schools) of the development of learning. . . . The VCR, the CD-ROM and now the Internet each represent a step in development that will eventually short-circuit the middle stage and its frustrating and psychologically dangerous dependence on adults and schooling. —**Seymour Papert,** in a 1995 essay he wrote for *Time* magazine titled "The Parent Trap"

Traditional Employment

The broader and perhaps more dramatic social impact of the hyperlearning revolution will be the large-scale displacement of traditional "employment" by a new form of human capitalism in which ownership of intellectual capital progressively replaces labor.—**Lewis J. Perelman,** of the Discovery Institute, in a 1994 *Wired* article titled "School's Out: The Hyperlearning Revolution Will Replace Public Education"

Many jobs are just never coming back. Blue-collar workers, secretaries, receptionists, clerical workers, sales clerks, bank tellers, telephone operators, librarians, wholesalers and middle managers are just a few of the many occupations destined for virtual extinction in the Digital Age.—**Jeremy Rifkin,** the author of the book *The End of Work,* quoted in a 1995 *Seattle Post-Intelligencer* article titled "Scholars Try to Measure the Impact"

Electronic communication technology will change, is changing, indeed, has changed (although some of us haven't noticed it) the nature of work, and indeed the need for work, in all advanced countries. . . . it may be difficult for most of us to realize that "work" is over, and that very few humans are needed any more. . . . Our old ideas and institutions will be smashed and paved over by the white-hot oozing asphalt of the Information Superhighway.—**Jim Dator,** in a 1994 speech titled "Does Religion Have a Future?"

Traditional Publishing

In the electronic age, the old notion of what is published will disappear. What is being written is going to be used in ways never imagined and writers and other creators are going to have to be aware of the dangers. . . . We are going to have to create a cultural workers' federation to marshal everyone's collective powers in the electronic age.—**Jonathan Tasini,** union president, quoted in a 1993 Reuters article headlined "Writers' Union Eyes Computer Lawsuit"

Traditional Wholesale and Retail Sales Structures

Many of today's warehouses, retail establishments, and last-mile delivery by housewives and other consumers, will not be needed. . . . Market-making businesses, like real-estate listings, travel agencies, and security and commodity brokerage may disappear. Staple items may be purchased online, and delivered via services like United Parcel or new local or manufacturer-owned delivery companies. Industrial and durable goods, where in-depth, comparative information and computer-based analysis tools will be most important, will also be examined and screened online. . . . The employment and resource allocation changes generated by online shopping during the coming century might be comparable to the shifts out of agriculture.—**Larry Press,** a professor of computer-information systems, in a 1994 article for *Communications of the ACM* titled "Commercialization of the Internet"

Merchants will find that they can dispense with sales floors and sales staff altogether and just maintain servers with databases. . . . Consumers might either "window shop" by remotely accessing such virtual stores, or they might delegate the task to software shopping agents that go out on the Net with shopping lists, inspect the specifications and prices of the merchandise on offer, and return with reports on the best available matches and prices. Closure of a sale can immediately trigger a delivery order at a warehouse, update an inventory database, and initiate an electronic money transfer. . . . The stock is bigger and the selection larger than in the mightiest big-box off-ramp superstore. The things that remain in physical form are warehouses . . . and delivery vehicles.—**William Mitchell,** in his 1994 book *City of Bits*

Traditional Schools

The telecomputer could revitalize public education by bringing the best teachers in the country to classrooms everywhere. More important, the telecomputer could encourage competition because it could make home schooling both feasible and attractive. To learn social skills, neighborhood children could gather in micro-schools run by parents, churches or other local institutions. The competition of home schooling would either destroy the public

school system or force it to become competitive with rival systems.—Writer
Roger Karraker, in a 1991 *Whole Earth Review* article headlined "Highways of
the Mind or Toll Roads Between Information Castles?"

With multi-media, there is no need for any human mediator or gatekeeper at
all. Education can finally be random access according to each learner's whim.
. . . the transformation of education is not far away. No more introductory
courses! No more sequences of courses! No more standardized tests of
sequentially-acquired data! No more standardized evaluations! Rather, everyone moves at her own pace down pathways of her own choosing, through
never-ending sequences of ever-opening doors . . . forever. How do you graduate—how can you be certified—who evaluates whom—when there is so
much more to explore? This is truly "continuing education" for eternity.
Clearly, the Classroom of 2010 will be everywhere and nowhere. . . . We teachers for the most part are dead ducks.—**Jim Dator,** in a 1993 speech for the
Seminar for Presidents of Community and Junior Colleges titled "The College
Classroom of the Year 2010"

The creation and transmission of knowledge will no longer move vertically,
from the top down. It will move horizontally, among many people, at a tremendous speed. This will undermine the foundation of every bureaucracy, including schools. . . . The very notion of traditional education will become obsolete.
Learning will not be based, as it is today, on mechanisms of selection and ex-
clusion. . . . Diplomas will disappear. Instead, people will get certificates (the
same way we get driver's licenses) to show potential employers that they have
specific skills, talents, or knowledge.—**Lewis Perelman,** quoted in a 1993 *Christian Science Monitor* article titled "Will Technology Alter Traditional Teaching?"

What follows from imagining a Knowledge Machine is a certainty that school
will either change very radically or simply collapse. It is predictable that the
education establishment cannot see farther than using new technologies to do
what it has always done in the past, teach the same curriculum. . . . The possibility of freely exploring worlds of knowledge calls into question the very idea
of an administered curriculum.—**Seymour Papert,** 1993

We will pay children to teach other children, adults to teach adults, and
children to teach adults as well. A system of royalties will be created on

information-server networks, like America Online, Prodigy, etc. As material is used, the provider earns money, and as they consume other's ideas they pay out. An electronic economic marketplace for ideas and knowledge will be created. Schools may well begin this, or perhaps be replaced by it.—Education technology consultant **Ed Lyell**, in a 1994 speech for the New York State Deans of Education

Home schooling will become more attractive when network resources become available, and this alternative will pressure schools to increase their quality or face widespread public rejection.—Education technology consultant **Bob Tinker**, in a 1995 white paper titled "The Future of Networking Technologies for Learning"

THEN THERE'S THE DEATH OF INTERNET ITSELF . . .
The scenario I'm playing with now is that the Internet might die. You can imagine a situation in which there's 200 million people on the Internet trying to send e-mail messages and the whole thing just grinds to a halt. Its own success just kills it. In the meantime, a telephone company steps in and offers e-mail for $5 a month, no traffic jams and it's reliable. I hope it doesn't happen, but it's a scenario one has to consider.—**Kevin Kelly** made this remark to a *London Observer* reporter in 1994, immediately after saying he'd been interviewed about the Internet so many times that, "usually I make something up."

AND, FINALLY, THE EXTINCTION OF THE HUMAN RACE
You end up with [robots] forming a cyberspace where entities try to outsmart each other by causing their way of thinking to be more pervasive. . . . The competitive pressure toward miniaturization will result in activity on the subatomic level. They'll transform matter in some way; it will no longer be matter as we know it. . . . I don't think humanity will last long under these conditions. . . . The robots will re-create us any number of times, whereas the original version of our world exists, at most, only once. Therefore, statistically speaking, it's much more likely we're living in a vast simulation than in the original version.—**Hans Moravec**, professor at Carnegie Mellon's Robotics Institute, quoted in a 1995 *Wired* article titled "Superhumanism: According to Hans Moravec, by 2040 Robots Will Become as Smart as We Are and then They'll Displace Us as the Dominant Form of Life on Earth"

8

Aristotle, Jefferson, Marx, and McLuhan

Predictors use historic perspective to make their points on issues

In their support or criticism of the new Internet technology in the period between 1990 and 1995, stakeholders and concerned critics pulled personalities from the past into play, referring to Aristotle, Socrates, Plato, the Sumerians, the Medicis, Jefferson, Paine, Madison, Franklin, Hamilton, King George, Marx, Thoreau, Verdi, T. S. Eliot, Rockefeller, Hitler, Stalin, and Einstein. For instance, in a 1994 article headlined "The Couch Potato Vote," *Wired* magazine reporter Evan Schwartz wrote, "Madison and Hamilton might retch at the vision of sofa spuds choosing to ratify or eradicate NAFTA with a click-click of their remote controls or a beep-beep of their touch-tone phones. . . . Thomas Jefferson might find electronic town meetings an absolute scream."

As knowledge communities, social webs, and communities of practice began to formally develop, people began learning how to create, organize, and access networked communications in ways that best suited their needs. It was evident that the emerging knowledge economy would be dependent upon trust. Linking the names of heroes of the past to the Internet was a way to build trust; linking the names of despots of the past was a way to tear it down.

Linking the Internet to successful networks of the past was a natural, and it was played out to the hilt in the "information highway" metaphor.

In their 1993 *Business Week* article "What's Arriving on the Information Highway?" Christopher Farrell and Michael Mandel projected: "The raw investment numbers understate the dynamism that will be unleashed by building the Information Superhighway. Much like the construction of the railroads in the 19th century, electricity networks in the 20th century, and the interstate highway system after World War II, the Information Superhighway will change the way we live at home and work. It will also open up opportunities for new goods and services dreamed up by the nation's innovators and entrepreneurs—cutting-edge ideas that spur economic growth."

The pundits and experts of the early 1990s drew parallels to the development and impact of "old" technologies and networks to illustrate their ideas and ideals regarding the new networked communications they saw sweeping the world into a new age—they mentioned Gutenberg's printing press, railroads, the automobile, radio, television, and the space program. For instance, in a 1995 article for *Wired* titled "Déjà Vu All Over Again," researcher Todd Lappin wrote, "Radio started out the same way. It was a truly interactive medium. It was user-dominated and user-controlled. But gradually, as the airwaves became popular, that precious interactivity was lost. . . . We are present at the creation of yet another great system whose worth will depend on the use we make of it. . . . Our job is to make sure that glorious potential doesn't get stuffed into yet another tired, old media box."

These predictors of the Internet's future—pro and con—drew historic parallels to the French Revolution, the Oklahoma land rush, the Wild West, Prohibition, and the Japanese attack at Pearl Harbor during World War II. For example, at a conference on "information warfare" in 1995, Ronald Gove of Science Applications International said, "We are facing an electronic Pearl Harbor. All you need is a PC, a modem, a little skill and a lot of patience, and you can do a lot of damage."

The predictors also quoted respected theorists of the past, including Vannevar Bush and Marshall McLuhan. In a 1995 *New York Times* interview titled "Present at the Creation; Startled at the Reality," science fiction novelist William Gibson said, "The present is more frightening than any imaginable future I might dream up. . . . If Marshall McLuhan were alive today, he'd have a nervous breakdown."

Canadian-born academic McLuhan was a communications technology expert whose influential books *The Gutenberg Galaxy* (a 1962 book in which he coined the phrase "global village"), *Understanding Media: The Extensions of Man* (1964), and *The Medium Is the Message: An Inventory of Effects* (1967) assured his place in both scholarly and pop-culture circles throughout the latter decades of the 20th century. He was a founder of modern media studies, and he was called in as a consultant by media and advertising interests including IBM, General Electric, and AT&T. He even appeared in the 1977 Woody Allen movie *Annie Hall,* in a brief humorous cameo as himself.

McLuhan painted technology as an extension of the human nervous system and said new technologies create new human perception. His work in deconstructivist criticism was on the cutting edge of its time. He emphasized the process over the product and the form over the content, explaining it in memorable aphorisms including, "The medium is the message." He was criticized by some of his contemporaries for publishing "derivative" work based on the ideas of others; some say this criticism was sparked by jealously over his public popularity. He died in 1980, but his influence lived on; *Wired* magazine named McLuhan its patron saint in its first issue in 1993.

In the early 1990s, predictors made comparisons in which the coming of networked communications somehow could be associated with the Khyber Pass, the Oregon Trail, and the Maginot Line. There were also many references to the Holy Grail. Following is a selection of quotes with historic ties in which the authors also make some sort of predictive statement about the future of the Internet.

IT'S ALL GREEK, OR ANCIENT, ANYWAY . . .

Network-based education has the benefits of mass production and mass distribution. It can also work in a tutorial mode so that it can deliver answers to a student's questions. It combines the best post-industrial production methods with the capacity to deliver education in the best Socratic tradition.—**Paul Strassman,** in his 1990 book The Business Value of Computers

Telecomputing . . . can become an agent of educational reform, a kind of Trojan horse that quietly but thoroughly rearranges classrooms.—Writer **Jacques Leslie,** in his 1993 *Wired* article "Kids Connecting"

Notions of property, value, ownership, and the nature of wealth itself are changing more fundamentally than at any time since the Sumerians first poked cuneiform into wet clay and called it stored grain. Only a very few people are aware of the enormity of this shift, and fewer of them are lawyers or public officials. Those who do see these changes must prepare responses for the legal and social confusion that will erupt as efforts to protect new forms of property with old methods become more obviously futile, and, as a consequence, more adamant.—EFF co-founder **John Perry Barlow,** in his 1994 *Wired* article "The Economy of Ideas"

The Internet demonstrates the possibility of a multilayered, heterogeneous, decentralized system in which the constituent communities organize themselves, run their local affairs, and pay their bills in many different ways. . . . These models will be debated, extended, and transformed. The fundamental questions of cyberspace's political economy will urgently be contested. Who plays, who pays, and how is this decided? How is trade to be conducted, and how is intellectual property to be managed and protected? What is the role of agents, and what sorts of regulation might these software slaves require? How should communities define their boundaries, and how might they maintain their norms within these boundaries? What are the legitimate forms of power? How might political discourse be constructed? These are questions worthy of an online Aristotle.—MIT's **William Mitchell,** in his 1994 book *City of Bits*

The country may be moving in the direction of purer democracy than anything the ancient Greeks envisioned. It promises to be a fiasco. Opinion polls and focus groups are Stone Age implements in the brave new world of interactivity just down the communications superhighway. Imagine an ongoing electronic plebiscite in which millions of Americans will be able to express their views on any public issue at a press of a button. Surely nothing could be a purer expression of democracy. Yet nothing would have a more paralyzing impact on representational government. . . . Now imagine the paralysis that would be induced if constituents could be polled instantly by an all-but-universal interactive system. No more guessing what the voters were thinking; presidents and lawmakers would have access to a permanent electrocardiogram, hooked up to the body politic.—ABC *Nightline* TV anchor **Ted Koppel,** quoted in Lawrence Grossman's 1994 book *The Electronic Republic*

The worldwide computer network—the electronic agora—subverts, displaces, and radically redefines our notions of gathering place, community, and urban life. . . . It will play as crucial a role in 21st-century urbanity as the centrally located, spatially bounded, architecturally celebrated agora did (according to Aristotle's "Politics") in the life of the Greek polis and in prototypical urban diagrams like that so lucidly traced out by the Milesians on their Ionian rock.—**William Mitchell,** in his 1994 book *City of Bits*

I don't think anyone would have said that Socrates, Plato, and Aristotle were addicted to the Agora. The computer nets are the modern Agora, serving a role similar to talk radio and tabloid journalism, but with more participation, less sensationalism, and more thinking between remarks.—A multi-user domain founder named **Barry Kort,** quoted in Sherry Turkle's 1995 book *Life on the Screen*

Interactive information technology has the potential to become the 21st century's electronic version of the meeting place on the hill near the Acropolis, where 2,500 years ago Athenian citizens assembled to govern themselves. The electronic republic cannot be as intimate or as deliberative as the face-to-face discussions and showing of hands in the ancient Athenians' open-air assemblies. But it is likely to extend government decision making from the few in the center of power to the many on the outside who may wish to participate.—**Lawrence Grossman,** in his 1995 book *The Electronic Republic*

Better and more efficient delivery of what already exists is what most media executives think and talk about in the context of being digital. But like the Trojan horse, the consequence of this gift will be surprising. Wholly new content will emerge from being digital, as will new players, new economic models, and a likely cottage industry of information and entertainment providers.—MIT visionary **Nicholas Negroponte,** in his 1995 book *Being Digital*

THE HORSELESS-CARRIAGE COMPARISON . . .
The future of computing lies in the development of new systems as radically different from today's offerings as the original PCs differed from their predecessors. The personal computer may turn out to be like the horseless carriage, whose effect on our expectations was greater than its impact on our lives. PC

users daydreamed of electronic cottages and global villages but settled for word processors and spreadsheets instead.—Futurist **Paul Saffo,** in a 1991 *New York Times* article headlined "Farewell PC—What's Next?"

Folks, I'm not holding my breath. The information superhighway may or may not become a truly transforming technology—the likes of the railroad, car or phone. But if it does, the event is many years, and perhaps decades, away. . . . In every age there are millions of people like me—who are vying to be last to have whatever is new and allegedly better. Who has time for all the multimedia, interactive mumbo-jumbo? We will join the wave of the future only after it laps gently onto shore. Let others enjoy the pleasures of digging their Model T out of the mud.—*Newsweek* essayist **Robert Samuelson,** in a 1993 column headlined "Lost on the Information Highway"

It is companies that shun the PC today in order to cater to the TV, consumer electronics, and telephone industries that will end up in luxury backwaters. . . . Just as the real action was not at Churchill Downs or the Peapack Hunt Club, but in Detroit [in the first years of the automobile], the real action today—the source of wealth and power—is not at Nintendo or Sega, Sony or QVC; it is in the scores of thousands of computer and software companies comprising the industrial fabric of the information age—the exhilarating new life after television.—Technology consultant **George Gilder,** in his 1994 book "Life After Television"

I don't think that the human race has seen anything in the long run that's going to impact it like the Web even since Henry Ford rolled the first Model T off the production line.—**Steve Adams,** of Online Systems Services, quoted in a 1994 *Denver Post* article headlined "Whole World Is Caught in the Web"

"JEFFERSONIAN" WAS THE PREFERRED STYLE . . .

The machines behind the network can tally national opinion not only instantaneously but with a level of delicate precision that would have left Jefferson breathless.—**Peter Huber,** senior fellow at the Manhattan Institute, in a 1992 *Forbes* article headlined "Telephone Democracy"

Crucial doubts remain as to whether the re-wiring of America will result in Jeffersonian networks promoting the openness, freedom, and diversity that is the true promise of this technology. In the worst case, we could wind up with networks that have the principal effect of fostering addiction to a new generation of electronic narcotics. . . . Their principal themes revolving around instant gratification through sex, violence, or sexual violence; their uses and content determined by mega-corporations pushing mindless consumption of things we don't need and aren't good for us. . . . The Jeffersonian ideal—a system that promotes grassroots democracy, diversity of users and manufacturers, true communications among the people, and all the dazzling goodies of home shopping, movies on demand, teleconferencing, and cheap, instant databases—is composed of high bandwidth, an open architecture, and distributed two-way switching. It's our choice to make. Let's not blow it.—EFF co-founder **Mitchell Kapor**, in his 1993 *Wired* article "Where Is the Digital Highway Really Heading?"

We have an opportunity to create media that would match the splendid ambitions of Franklin with his public libraries and his mail system, and Jefferson and Madison with their determination to arm democracy with the power knowledge gives. We could offer children, yes even poor children in poor districts, a real opportunity to control the screen, for once.—Author **William Gibson**, in a 1993 speech for the National Academy of Sciences Convocation on Technology and Education

Diversity will maximize profit only by assuring the largest possible customer base. Since when does mass appeal mean the flourishing of free, autonomous, yet interdependent individuals that is the hallmark of a "Jeffersonian democracy?" In his National Press Club speech, the vice president [Al Gore] was closer to the truth: "We'll turn from consumers into providers. In a way, this change represents a kind of empowerment." . . . Unfortunately, the vice president went on to liken individuals under the "communications revolution" to mere instrumental sources of informational "added value" to the economy, like factory workers. That's precisely the wrong analogy if networks are to form the basis of an electronic "Jeffersonian democracy." . . . The NII would scarcely be worth building if it offered no more than 500 channels of MTV, no matter how holographic, ambient, and jacked in to the gills. Its real payoff, its visionary promise, would be the possibility of an "Athens without slaves" or a "Jeffersonian democracy" in which people can provide information as easily as they consume it. A networked world offers the possibility of many-to-many communication, permitting widely separated individuals to bind themselves into collectives.—New York technology lawyer James Cappio, in a 1994 *Wired* article titled "Bad Attitude: Business as Usual on the Infobahn"

As the 20th century draws to a close, the idea of a virtual campus— paralleling or perhaps replacing the physical one—seems increasingly plausible. If a latter-day Jefferson were to lay out an ideal educational community for the third millennium, she might site it in cyberspace.—William Mitchell, in his 1994 book *City of Bits*

Never before has a group of citizens with such global awareness, depth of experience, media sophistication, and healthy skepticism been handed such a massive opportunity. Without any official proclamations, we're already in the information age. We're no longer at war, and we live in a networked economy. The left wing is dead. The right wing is dead. Ideology is dead. In place of the stale, 19th-century pre-cyber age ideologies that still provide coinage for "the system" (how could anyone still be proud to be identified as a socialist or a Jeffersonian or a libertarian?), proto-movements are beginning to form to tackle the far more radical politics of cyberspace.—Mark Stahlman, president of New Media Associates, in a 1994 *Wired* article headlined "Just Say No—To Cybercrats and Digital Control Freaks"

There is a fundamental problem with a system that requires, through technology, payment for every access to a particular expression. It defeats the original Jeffersonian purpose of seeing that ideas were available to everyone regardless of their economic station. I am not comfortable with a model that will restrict inquiry to the wealthy.—John Perry Barlow, in a 1994 *Wired* article titled "The Economy of Ideas"

JON KATZ ON THOMAS PAINE . . .
The new media (computers, cable, and the Internet) can and should adopt him. . . . [Thomas] Paine does have a legacy, a place where his values prosper and are validated millions of times a day: the Internet. There, his ideas about communications, media ethics, the universal connections between people, and the free flow of honest opinion are all relevant again, visible every time one modem shakes hands with another. . . .

Today's media are what the Net should never become—but will surely evolve into if it fails to develop, articulate, fight fiercely for, and maintain a value system other than expanded memory, whiz-bang toys, and money. . . . The political, economic, and social implications of an interconnected global medium are enormous, making plausible [Thomas] Paine's belief in the "universal citizen." . . .

[Cyberspace] would be Paine's home now. . . . If Paine would feel at home there, he would also fight to protect this nascent medium. . . . He would spot commercialization as Danger Number One. He believed in a press that was not monopolistic but filled, as it was in his time, with individual voices; one that was cheap, accessible, fiercely outspoken. He believed that media like the Net—many citizens talking to many other citizens—were essential to free government. . . . he'd have lots to say about the so-called information highway and the government's alleged role in shaping it.—Media commentator **Jon Katz,** in a 1995 essay for *Wired* headlined "The Age of Paine . . . He Should Be Resurrected as the Moral Father of the Internet"

GOODBYE TO GUTENBERG . . .

Electronic text processing marks the next major shift in information technology after the development of the printed book. It promises (or threatens) to produce effects on our culture, particularly our literature, education, criticism and scholarship, just as radical as those produced by Gutenberg's movable type.—**George P. Landow,** in his 1991 book *Hypertext*

At first, the bicycle was a silly contraption; the automobile, a noisy intruder, the pocket calculator, a threat to the study of mathematics; the radio, an end to literacy. But then something happens. Over time, these machines find a place in our everyday lives. . . . A new generation grows up with them, changing and humanizing them. In short, playing with them. The telephone was a major advance in two-way communication. But at first, even it was denounced as nothing more than a nuisance. People were made uncomfortable and awkward by this mechanical invader in their homes. Eventually, though, men and women realized they were not just getting a new machine, they were learning a new kind of communication. . . . As it flourished, its own special expression, tricks, etiquette, and culture developed. . . . As I write, a newer form of communication—electronic mail, e-mail, is undergoing the same sort of

process: establishing its own rules and habits. . . . The information highway
will transform our culture as dramatically as Gutenberg's press did the Middle Ages.—Microsoft CEO **Bill Gates,** in his 1995 book *The Road Ahead*

TELEGRAPH, TELEPHONE, RADIO, AND TELEVISION . . .

A lot of this "too hard to use" stuff will go away. Radio was so messy for the
first 20 years, it wasn't funny. Cars ditto—you had to be a mechanic to drive
one.—**Carl Malamud,** of Internet Multicasting Service, quoted in a 1993 *Fortune* magazine article headlined "Boomtime on the New Frontier"

While no one can predict with certainty the future of interactive television,
the thrust of history and the force of logic suggests that if the marketers are at
the helm, it is bound to end up looking more like QVC than the Internet. . . .
The history of broadcast has its own inexorable logic, a logic based on selling
an audience about which progressively more can be known. The $64,000
Question is whether interactive television will reinforce that logic, taking it to
the extreme of one-consumer/one-commercial, or shatter it, creating a decentralized network of citizen-broadcasters, each with his own vision of the future. The answer is television history in the making.—Writer **Adam Fisher,** in
the 1993 *Wired* article "Do Nielsen's Broadcast Ratings Have a Future in the
Coming Interactive Age?"

The telegraph is the ancestor of just about everything modern we know today,
and this year it's 150 years old. I think it's fitting that this is also the first year
that the Internet fully sheds its experimental status and takes its place as a
fledgling medium along with print and TV. Like the telegraph, the Internet
will surely spawn all kinds of inventions and new ways of doing things, things
we can't imagine today. And someday, strangely enough, our descendants will
look back at 1994, the year of the birth of the Internet-as-medium, as the old
days and wonder how in the world we ever got by with such primitive technology!—**Ken McCarthy,** founding publisher of the *Internet Gazette,* in his
1994 speech to the San Francisco conference "Why the Web and Why NOW"

Thomas Edison thought his new invention, the phonograph, would be used
as a dictating machine. Alexander Graham Bell looked upon his telephone as
an entertainment medium. The *New York Times* predicted that television was

unlikely to catch on as radio did. . . . We are in the midst of a telecommunications revolution whose characteristics are yet to be clearly defined. There will be no shortage of channels into and out of the home. In fact, in an age of video dial tone and digital transmission, there may no longer be channels—only unlimited bits of information and data to be translated into any format one wants to call up.—**Lawrence Grossman**, in his 1995 book *The Electronic Republic*

"For the first time in human history we have available to us the ability to communicate simultaneously with millions of our fellowmen, to furnish entertainment, instruction, widening vision of national problems and national events. An obligation rests on us to see that it is devoted to real service and to develop the material that is transmitted into that which is really worthwhile." [Who uttered the preceding quote?] Mitch Kapor? Newt Gingrich? Al Gore? Alvin Toffler? Nope. Herbert Hoover, speaking in 1924 as the Secretary of Commerce. And the "great system"? Not the Internet. Nor the Infobahn. It was radio. . . . Today's Next Big Something is so wrapped in hype it's tough to see what's really going on. . . . By exchanging information, we become free. Blah, blah, blah. But what if conventional wisdom is wrong? What if the crystal-ball narrative doesn't turn out as planned? What if, a decade or so from now, we wake up to find that the digisphere has been overrun by swarms of inane mass marketers? . . . Radio started out the same way. It was a truly interactive medium. It was user-dominated and user-controlled. But gradually, as the airwaves became popular, that precious interactivity was lost. . . . We are present at the creation of yet another great system whose worth will depend on the use we make of it. . . . Our job is to make sure that glorious potential doesn't get stuffed into yet another tired, old media box.—Writer **Todd Lappin**, in his 1995 *Wired* article headlined "Déjà Vu All Over Again"

In the 1970s, cable was promised as a new vehicle to provide education to the home. But today, far more shows center around the theme of murder than of education; corporations profit from "lowest-common-denominator entertainment" and have sidestepped cable's educational potential. We risk repeating this pattern with top-down implementation of the NII.—**Frank Odasz**, director of Big Sky Telegraph, in his article "Issues in the Development of Community Cooperative Networks," published in *Public Access to the Internet*

MCLUHAN IS GRAND MARSHALL . . .

According to McLuhan, the tedium of print culture was unrelieved until the onset of widespread use of radio and television in the 20th century. [Walter] Ong characterizes the changes in language and thought growing from the radio and television broadcasts as a secondary orality. This discussion posits the existence of a tertiary form of orality, exhibited in computer-mediated communication (CMC) systems. This . . . occurs in real-time computer conferencing systems and in asynchronous computer bulletin board systems. Although based on text, the discourse in these computer-mediated forums exhibits many qualities of an oral culture. The existence of this text-based orality may imply that discourse need not be based upon sound in order to have oral characteristics. . . . Just as the earlier technologies (the Greek alphabet, the printing press) changed the way people communicated and thought, so, too, does CMC.—Technology editor/writer **John December,** in his 1993 research paper "Characteristics of Oral Culture in Discourse on the Net"

It is unlikely that Michael Heim, David Bennahum, Don Langham, or any member of the first generation of cybernauts will be able to accurately describe or assess the realm of the new technology; as Marshall McLuhan has pointed out, we are blinded by our own cultural upbringing, and the realm of cyberspace is certainly one that cannot be inhabited for at least another full generation.—Researcher **Michael Doherty,** in a 1994 *Computer-Mediated Communication* magazine article titled "MOO as Tool, MOO as Realm: A Response to Don Langham"

We are trying (as Marshall McLuhan said) to retribalize. And the computer is playing a central role. . . . These shifts raise many questions. What will computer-mediated communication do to our commitment to other people? Will it satisfy our needs for connection and social participation, or will it further undermine fragile relationships? What kind of responsibility and accountability will we assume for our virtual actions?—Researcher **Sherry Turkle,** in her 1995 book *Life on the Screen*

With digital transmission, signal compression, fiber optic transmissions, and expanded use of the electronic spectrum, information of all kinds in all forms will pour in and out, limited only by people's ability to pay. It will travel via cable lines and telephone lines, through the air, and directly to and from satellites. This interactive telecommunications revolution is already in the process of permanently and profoundly changing our nation's political system. "As the speed of information increases," McLuhan wrote, "the tendency is for politics to move away from representation and delegation of constituents toward immediate involvement of the entire community in the central acts of decision."—**Lawrence Grossman,** in his 1995 book *The Electronic Republic*

Marshall McLuhan's famous statement—"We shape our tools, and thereafter our tools shape us"—is . . . true for the future. . . . The most potent technology transforming the present . . . is the vast array of electronic communications technologies which are now being widely touted as composing the Information Superhighway. . . . Many once-separate and expensive technologies are being woven together into a gigantic, global, and comparatively inexpensive information network which, among other things, is destroying the necessity of traveling to a single centralized location to work, or to trade, or be entertained, or even to govern. It is now increasingly possible, and preferable, to telework, to telemarket, to teleview, and to telegovern. It thus is no longer necessary, nor desirable, anywhere in the world to continue to create huge urban centers. . . . It can all come to you.—Futurist **Jim Dator,** in a 1995 speech for the International Conference on Development, Ethics, and the Environment, in Kuala Lumpur

IN SEARCH OF THE HOLY GRAIL . . .
The Holy Grail in the consumer market will be to bring new applications into the home. Chief amongst those applications will be software that will provide two-way communications through a new device we call the TV/PC.—**Bill Gates,** in a 1993 *Red Herring* magazine article headlined "Bill Gates: An American Gladiator in the Digital Arena"

The perfectly friendly interface is too often seen as the Holy Grail of online service designers. It's the application, stupid! It doesn't matter how cute the icons are in my spreadsheet software—I'm not going to use it to balance my checkbook. . . . If the online industry is going to attract a lot of new users, we're going to have to spend less time inventing new features and more time figuring out the answer to the fundamental question of defining something of value that people need and want. . . . Perhaps the thing online services could do today to make the technology more user-friendly is hire a few less programmers and a lot more guides, teachers and facilitators to bridge the gap between the complex resources and people who currently find them arcane. —**Lisa Kimball,** a partner in Metasystems Design Group, in a 1993 *Wall Street Journal* article headlined "Technology: Two Dozen Users and Analysts Examine the Potential—and Shortcomings—of Networks"

What would be the ideal in a few years if everything happened right? There would be a single cross-platform plug-in object technology where you could slot components in. It's sort of the Holy Grail right now. I'm skeptical that there will be enough cooperation between vendors to make it happen.—**Marc Andreessen,** in a 1995 online interview with Barry Phillips

AND NOW, A LITTLE BIT OF EVERYTHING . . .

From 1992–1993—Romans, Rural Electrification, and Space

We can expect the speed of "real time" to help us project into cyberspace some of our dearest phantasms, some of our worst monsters. This power of revelation and embodiment will be felt by many to be the utmost obscenity. Let's not forget that both Hitler and Stalin are known for having banned the publication of fairy tales. . . . Freedom of imagination is feared by most powers.—Writer **Nicole Stenger,** in her 1992 essay "Mind Is a Leaking Rainbow," included in the book *Cyberspace: First Steps*

I do not believe there will be a Bit Police. The FCC is too smart. Its mandate is to see advanced information and entertainment-service proliferate in the public interest. There is simply no way to limit the freedom of bit radiation any more than the Romans could stop Christianity, even though a few brave and early data broadcasters will be eaten by the Washington lions in the process.—**Nicholas Negroponte,** in his 1993 *Wired* column headlined "The Bit Police: Will the FCC Regulate Licenses to Radiate Bits?"

I believe that the national adventure you now propose is of quite extraordinary importance. Historians of the future—provided good dreams prevail—will view this as having been far more crucial to the survival of democracy in the United States than rural electrification or the space program.—**William Gibson,** in his 1993 speech for the National Academy of Sciences Convocation on Technology and Education

John Sculley, one of the network's foremost proponents, is even calling on Clinton to set a goal for its construction the way President Kennedy called for putting a man on the moon. Such a network could cost at least $200 billion to build.—*USA Today* reporter **Kevin Maney,** reflecting the remarks of Apple Computer CEO John Sculley, in a 1993 article titled "A New Superhighway"

The American Revolution was the first war fought, in part, through public opinion in the newspapers, and Ben Franklin was the first media-savvy lobbyist to employ techniques of disinformation. For the next 200 or so years, the media have been able to behave in a basically monopolistic way. They have treated information the way John D. Rockefeller treated oil—as a commodity, in which the distribution network, rather than product quality, is of primary importance. But once people can get the raw data themselves, that monopoly ends. . . . I will be able, for example, to view any public meeting of Congress over the Net. And I will have artificial intelligence agents roaming the databases, downloading stuff I am interested in, and assembling for me a front page, or a nightly news show, that addresses my interests. I'll have the 12 top stories that I want, I'll have short summaries available, and I'll be able to double-click for more detail. How will Peter Jennings or MacNeil-Lehrer or a newspaper compete with that?—Author **Michael Crichton,** in his 1993 *Wired* article headlined "Mediasaurus"

We should have learning centers, neighborhood electronic cottages. In a sense it's going back to the pioneer days where you had small schools with students of different ages and just one or two teachers overseeing them and teaching many subjects.—Education technology expert **Ed Lyell,** quoted in a 1993 *Wired* article headlined "Man with a Plan"

From 1994—Socrates, Willie Sutton, Adam Smith, and Disney

Our concept of cyberspace, cyber-culture, and cyber-everything is, more than we care to realize, a European idea, rooted in Deuteronomy, Socrates, Galileo, Jefferson, Edison, Jobs, Wozniak, glasnost, perestroika, and the United Federation of Planets. . . . During [a two-week trip to Hong Kong, Shenzhen, and Shanghai during September 1993] I tried to get some sense of how the Chinese perceived the influence of technology—particularly digital technology—on their culture. The answer is that this issue hasn't occurred to the Chinese yet, and probably never will, because it basically stems from a Western, post-Enlightenment perspective.—Writer **Neal Stephenson, 1994** *Wired* article headlined "In the Kingdom of Mao Bell"

The mistakes of the French Revolution are being repeated. On the Internet, it is possible to publish anonymously or pseudonymously and more widely than ever before. As a result, the responsibility for organizing information shifts from the writer to the reader. How can you know what to believe? This is a time of great danger.—**Carla Hesse,** Berkeley history professor, quoted in a 1994 *New York Times* article titled "The Rise and Swift Fall of Cyberliteracy"

Back in the 1960s, when photocopying first became commercially successful . . . People adjusted the way they thought about things, and people really didn't go reproduce their books on Xerox machines. They certainly did copy parts of them. But it didn't kill the book industry. . . . There are ways to get compensated for things that aren't part of the model that exists today, but will appear as part of the new network environment.—**John Warnock,** founder of Adobe Systems, in a 1994 *Wired* article headlined "Scriptwriter"

Personal terminals, for making and receiving payments anywhere, could be integrated with laptop or palmtop computers or could be specialized wallet-sized devices. . . . Cash money and associated transaction points may soon disappear entirely. Today's Willie Suttons are learning to crack computer security, not safes.—**William Mitchell,** in his 1994 book *City of Bits*

Schools are one of the principal barriers to the growth of not only this new industry, but the whole world economy. Replacing the bureaucratic empire of educational institutions with a high-tech commercial industry will pull the cork out of the knowledge-age bottleneck. . . . The institution of contemporary, "public" education is a 19th-century innovation designed as a worker-factory for an industrial economy. Both have as much utility in today's modern economy of advanced information technology as the Conestoga wagon or blacksmith shop.—**Lewis J. Perelman,** in a 1994 *Wired* article headlined "School's Out: The Hyperlearning Revolution Will Replace Public Education"

Nations that seek to remain economically competitive and to provide high living standards for their citizens will race to embark on their National Information Infrastructure projects as, in the past, they have invested in their ports and shipping fleets, railroad networks, and highway systems. And as they do so, they will have to resolve fundamental questions about the political economy of cyberspace. . . . Democratic ideals (and the lessons of the telephone system) suggest that they should strive to provide universal access—affordable, ubiquitously present, high-bandwidth service to all their citizens. . . . If the infrastructure is to encourage national coherence rather than a new kind of balkanization, then its development must be guided by policies and standards that assure interoperability between all the subnetworks of the national system.—**William Mitchell,** in his 1994 book *City of Bits*

Computers multiply data; in fact, one study indicated that data would double 19 times between 1990 and 2000. How will anyone be able to find the information needed in this huge haystack? The world is already choking on data. . . . One might add: Where is the information we have lost in data?—Technology consultant **George Gilder,** in his 1994 book *Life After Television*

The Sixteenth Amendment, passed in 1913 to empower the federal government to collect a tax on incomes, has since fueled the growth of a bloated leviathan, unleashing unchecked inquisitorial powers that are used to pry into the most intimate details of our lives. Not too long after our incomes follow our economic activity into Cyberspace, this inquisition is going to end. Encryption is to the Information Revolution what the Atlantic Ocean was to the American Revolution. It will render tax authorities as impotent in projecting their power as the ocean crossing did to King George.—DigitaLiberty founder **Bill Frezza,** in a 1994 statement on the organization's site titled "The Crucible of Radical Capitalism: How the Information Revolution Will Transform the Politics of Power"

Computer networks, it is held, are instruments of liberty that allow people to communicate laterally, thereby breaking down the hierarchies of governments and corporations alike. The resulting vision is actually similar to that of Adam Smith, who thought of the market as a vast network of artisans and entrepreneurs and who had little or no inkling of the large, bureaucratic corporation. . . . The actual evidence, such as it is, points largely in the other direction: computer networks decentralize organization (in the sense of operational decision-making) while simultaneously increasing the power of corporate central management.—Information studies professor **Phil Agre,** in a 1994 article for *The Network Observer* headlined "The New Politics of Technology in the U.S."

To appreciate tomorrow's multimedia networks, don't look to the Bob Met-calfes, Ted Nelsons, and Vint Cerfs for ideas and inspiration. Those techno-wonks won't set the agenda; the Paleys, Sarnoffs, and Disneys of the world will. The economics of advertising, promotion, and sponsorship—more than the technologies of teraflops, bandwidth, and GUI—will shape the virtual re-alities we may soon inhabit. Wherever there are audiences, there will be ad-vertisers. As media evolve, so do audiences. Time and geography—more than human nature—separate the captive crowds at the Roman Colosseum from user lists on the Internet.—MIT Media Lab Fellow **Michael Schrage,** in a 1994 *Wired* article titled "Is Advertising Dead?"

From 1995—Nixon, Tocqueville, Einstein, and the Edsel
The idea of both fax and electronic mail goes back about a hundred years. In an 1863 manuscript, "Paris in the 20th Century," found and published for the first time in 1994, Jules Verne wrote, "Photo-telegraphy allowed any writ-ing, signature or illustration to be sent far away, and any contract to be signed at a distance of [20,000 km]. Every house was Wired." . . . When e-mail started during the middle and late [19]60s, relatively few people were computer liter-ate. . . . But today, with computer ubiquity, the advantages of e-mail are over-whelming, as evidenced by its skyrocketing use. . . . In the next millennium e-mail (by no means limited to ASCII) will be the dominant interpersonal telecommunications medium, approaching if not overshadowing voice within the next 15 years. We will all be using e-mail.—**Nicholas Negroponte,** in his 1995 book *Being Digital*

When Richard Nixon taped conversations in the Oval Office, he felt sure that only he and historians in the distant future would have access to them. I want to make sure that I can shut my agent off. I even want a periodic check on whether the agent holds any sensitive information. Sort of a security clearance for my agent.—**Sherry Turkle,** in her 1995 book *Life on the Screen*

[The cyberspace software race has begun.] It's like the land rush in Oklahoma. The best spot in the valley goes to the one who gets there first.—Oracle CEO **Larry Ellison,** quoted in a 1995 *Business Week* article titled "Crafting Software That Will Let You Build a Business Out There"

The telephone eroded the art of writing letters. Television cut into neighbor-hood cinemas. MTV and superstars weakened amateur musicians and hometown bands. The car destroyed urban trolley systems; interstate highways dev-astated passenger rail service; and airliners wiped out passenger ships. What is most at risk from wide-area networks? Our library system.—**Clifford Stoll,** in his 1995 book *Silicon Snake Oil*

Giuseppe Verdi once said, "Looking back at the past is a real sign of progress." In my opinion, futurologists such as Alvin Toffler tend to overemphasize the thread of technological determinism in history and then project it into the fu-ture. The technologies of transmission—writing systems, printing presses, and computers—do not necessarily drive change in a predictably specific di-rection. . . . Each technical step forward means a compensating step backward in our mind-sets.—French theoretician **Regis Debray,** quoted in a 1995 *Wired* article headlined "Revolution in the Revolution"

Upon leaving the White House in 1961, Dwight Eisenhower dubbed the military-industrial complex "a threat to democracy." He sure knew what he was talking about, since he helped build it up in the first place. But comes 1995, at the very moment that a military-informational complex is taking shape with some American political leaders, most prominently Ross Perot and Newt Gingrich, talking about "virtual democracy" in a spirit reminiscent of fundamentalist mysticism, how not to feel alarmed? How not to see the outlines of cybernetics turned into a social policy?—French theoretician **Paul Virilio,** quoted in a 1995 *Wired* article headlined "Speed and Information: Cyberspace Alarm!"

Grassroots democracy will be expressed by individuals sitting quietly in their kitchens and living rooms. . . . Such a scenario for the future fulfills the prophecy made last century by France's Tocqueville and Britain's Lord Bryce, two of history's more astute observers of the American democratic process: "By whatever political laws men are governed in the age of equality," Toc-queville wrote, "it may be foreseen that faith in public opinion will become for them a species of religion, and the majority of its ministering prophet." Simi-larly, Lord Bryce predicted a time in America when "public opinion would not only reign but govern," and the will of the majority would "become ascertain-able at all times, and without the need of its passing through a body of repre-sentatives, possibly even without the need of voting machinery at all." For the American people, that time will soon be at hand. For many it is already here.—**Lawrence Grossman,** in his 1995 book *The Electronic Republic*

Surely storage and delivery systems will improve to where a movie could be sent to a home computer to be viewed at leisure. Surely the electronic book of the future will be wieldier than a laptop computer. Surely badly designed computers will go the way of the Edsel.—Book reviewer **Christopher Lehmann-Haupt** disagrees with Clifford Stoll in a 1995 *New York Times* column about Stoll's book *Silicon Snake Oil*

Here lies a new and major risk for humanity stemming from multimedia and computers. Albert Einstein, in fact, had already prophesized as much in the 1950s, when talking about "the second bomb." The electronic bomb, after the atomic one. A bomb whereby real-time interaction would be to information what radioactivity is to energy. The disintegration then will not merely affect the particles of matter, but also the very people of which our societies consist. . . . One may surmise that, just as the emergence of the atomic bomb made very quickly the elaboration of a policy of military dissuasion imperative in order to avoid a nuclear catastrophe, the information bomb will also need a new form of dissuasion adapted to the 21st century. This shall be a societal form of dissuasion to counter the damage caused by the explosion of unlimited information.—**Paul Virilio**, quoted in a 1995 *Wired* article headlined "Speed and Information: Cyberspace Alarm!"

France's patriarchs are outraged. Europeans' love affair with Europe is threatened by this cultural flirtation with the Americans, and they are determined to put a stop to it. . . . The hype over interactive television, video-on-demand, and music delivered over the Internet, only strengthens the French resolve. If they do not take a stand now, they reckon, it will soon be too late. But, on the contrary, it's too late already. . . . In an age of interactive media, cultural quotas will prove at least as self-defeating—and if anything, useless—as the Maginot Line, France's last great attempt to wall itself off from invaders. . . . If Europe falls even further behind on that highway, it will no longer have to worry about its cultures, for it will have effectively put them all in a museum. As Molière once said: "Nearly all men die of their remedies, and not of their illnesses."—Economist correspondent **John Andrews**, in his 1995 *Wired* magazine article "Culture Wars"

Someone, probably Marx, made the observation that emerging classes tend to envision utopia in their own image. Small wonder then that here in the heart of the information economy the dream that seizes the imagination of our rising cyber class of entrepreneurs and code-warriors is one of empowerment and autonomy through greater information and technology. It's not a bad dream, really, although like most utopian visions it hinges on a certain mode of behavior becoming universalized—in this case computer literacy, gadget acquisition, and a voracious appetite for ersatz reality. If the Net truly does become the cultural glue holding the emerging global village together, the pressure toward such behavior will become relentless. Perhaps it will be fitting justice if the catalysts of the new digital politics are ultimately forced by the logic of their political ideals to become online Dr. Frankensteins battling their own creation run amuck.—Writer/editor **Jay Kinney,** in a 1995 *Wired* essay headlined "'Anarcho-Emergenist-Republicans': Is There a New Politics Emerging in the Net/Cyberspace/Digital Culture?"

There is something in us, possibly our own DNA, that recalls animistic ritual—the fear of night, of the gods who come back in different form on the big screen. If somebody comes along who's a persuasive demagogue and really commands the wires, commands the codes, and is an irresistible presence, I can imagine some dangerous scenarios. I can also imagine a spiritual leader tapping in and turning those same desires in the other direction. So it's a gamble. Electronic Church or electronic Reich? It could be either.—*Gutenberg Elegies* author **Sven Birkerts,** quoted in a 1995 *Wired* article headlined "Digital Refusenik"

I keep coming back to the image of old Ben Franklin and his printing press. Franklin understood that the British Empire was a dinosaur. Its bandwidth was no longer sufficient to support the extent of its body. So he used the innovative medium of his time to create bandwidth, thus setting into motion a form of social organization which could move faster and plan smarter than its obsolete competitor. So today we have the Net, the last accidentally uncensored mass medium in existence. Is it a toy of the rich and the ivory tower, or is it potent? . . . Will we allow ourselves to be possessed by the vision of a Net whose purpose is to help create and support HEROES? Or will we dismiss it all with a keystroke and get back to the REAL FUN STUFF on alt.flame.Joe.schmuck.the.world's.greatest.poophead?—Internet pioneer **Steve Crocker,** quoted in a 1995 *Computerworld* article headlined "Newbie Bashing"—it was an excerpt from J. C. Herz's book *Surfing on the Internet*

Supporters Crow About "500 Channels!" Everyone Warns About "Infoglut"

A breathless bromide about a video wonderland is bandied about, while information overload looms larger than ever

While *information superhighway* was a popular catchphrase for those wishing to describe the potential for a digital network communications grid in the early 1990s, any study of what was being written and spoken at the time turns up a second chant that became nearly as popular: *500 channels.*

Promoters of networked digital communications wanted everyone to know they could look forward to having "the entire Library of Congress" (another often-used Internet-age point of reference) at their fingertips, yet simultaneously even the most supportive backers of the rapid proliferation of the Internet warned that people were about to be swept away by untold amounts of data. *Infoglut* and *information overload* were the most popular shorthand references for describing the problem.

How could "500 channels" (in TV terms a high number, but actually an extremely low estimate that was not accurate in terms of the overall future of networked digital communications) and the realities of the incredible new communications network do anything but lead to infoglut? Here, we look at what led to all of this, and what people were saying about 500 channels and information overload as the Internet took off.

TELEPHONE AND CABLE COMPANIES, COMPUTER INDUSTRY LOOK AT CONVERGENCE

In the early 1990s, there was confusion and conflict over the type of technology that would ultimately be considered the best way to deliver interactive digital information to homes and businesses nationally and worldwide—who would build the superhighway? Included in the fray were the networking companies, telephone companies, cable companies, digital broadcast satellite (DBS) operators, and even the power companies. All had experience with networks leading to homes and offices, so they were positioned to be the prime providers of the delivery of digital information and to reap the expected profits to come.

At the time, DBS operators were just getting off the ground, and they were letting everyone know they had the capability to send their customers information on 500 channels. In the early 1990s, cable customers were generally receiving only several dozen channels, and many homes had not yet been wired for cable. Of course, the power companies and telephone companies all had existing lines to most homes, but in order to be able to deliver the Information Superhighway to these homes they would have to launch a multi-billion-dollar upgrade of those lines or initiate the swift development of better wireless communications technologies.

Most U.S. citizens of this era had lived most of their lives with only three or four television channels available in their homes—the typical number received in the 1950s, 1960s, and 1970s. And the broadcast day on those three or four channels would generally run from station sign-on at 6 A.M. to sign-off at midnight, with no nighttime programming of any kind. The 500-channels-24-hours chant was aimed at convincing these people to buy into the idea of digital, networked communications. Imagine going from three or four up to 500!

By 1993, telephone and cable companies began to form operating agreements and convergence became the buzzword as alliances between telephone and cable and the computer industry developed. The publicity surrounding the moves included the possibility of being able to offer 500 channels, which TCI President John Malone and others estimated would be possible with digital compression.

THE GOOD, BAD, AND UGLY OF 500 CHANNELS

Both supporters and critics of the rapidly expanding digital communications technologies used the phrase *500 channels* to make their points. Futurist Alvin Toffler said in a 1993 *Wired* magazine article headlined "Shock Wave

(Anti)Warrior": "The cultural reality that seems not very far off is 500 or 1,000 channels of television bringing in images from Fiji or Kazakhstan, automatically translated into my own language, carrying along ideologies and religions that blow the mind, and that create in every country a configurative culture, in which elements have been adopted from elsewhere in the world."

But for every positive statement about 500 channels there were many nasty remarks. Equating the new digital information age with the disappointing medium known as television was not likely to bolster its image.

A band named Choking Victim came out with an early 1990s song titled "500 Channels" that outlines a world with 500 channels of "day-dream stimulation"; the lyrics go on to predict this will cause people to resent their lives and "raise their expectations."

And New York lawyer James Cappio wrote in a 1995 *Wired* magazine article titled "Bad Attitude: Business as Usual on the Infobahn": "The NII would scarcely be worth building if it offered no more than 500 channels of MTV, no matter how holographic, ambient, and jacked in to the gills. Its real payoff, its visionary promise, would be the possibility of an 'Athens without slaves' or a 'Jeffersonian democracy' in which people can provide information as easily as they consume it. A networked world offers the possibility of many-to-many communication, permitting widely separated individuals to bind themselves into collectives."

But Vanderbilt University economists and Donna Hoffman and Thomas Novak saw beyond the "500-channel" hype to the point of the new information network, outlining their vision in a scholarly paper titled "Commercializing the Information Superhighway," published in a 1994 Vanderbilt business school publication. "Using a computer program like Mosaic to 'net surf' radically demonstrates how different the Internet is from the passive-interactive-multimedia vision of 500 channels on your TV set," they wrote. "The experience is social, full of information and driven by curiosity. And the Internet already offers more than 10,000 actively interactive channels. The millions of current Internet users are unlikely to be satisfied with a vision of the Information Superhighway that is anything less than what already exists."

And Mosaic creator and Netscape founder Marc Andreessen, in a 1994 interview with the Knight Ridder News Service, said the Internet would gobble up television. "The telephone and cable companies are developing their own computer systems to deliver programming, such as 500 channels of television, video on demand, and interactive games," he said. "During that time, the Internet, which already connects 20 million users, will continue to grow at an estimated 10 percent each month. By the time the new fiber-and-coaxial networks are built, the thinking goes, many people will be used to getting data from Mosaic. As a result, they'll demand that the software also control what comes into the home via the cable systems."

WHAT THEY WERE SAYING ABOUT THE "500 CHANNELS"

Following is a selection of a dozen statements made by stakeholders and skeptics between 1990 and 1995 in which the "500 channels" phrase comes into play in an intriguing or revealing manner.

We have to figure out what "universal access" means. . . . I don't think it means 24 hours of access to 500 channels of mud wrestling.—**Mike Nelson,** senior adviser at the White House Office of Science and Technology Policy, quoted in a 1994 *Business Week* article headlined "From Internet to Infobahn"

The entertainment/telephony/cable (ETC) community has three things they want to sell you: movies, games, and home shopping. Basically they're going to do it by giving you 500 channels of cable, with maybe a very small back channel that allows you to vote "yes" or "no," or purchase something. . . . [The government's] role is to provide just enough standardization—in just the right places—to let market forces speak in a competitive environment, but on the other hand, to keep people from being "locked out" by the competition.— Networking pioneer **Leonard Kleinrock,** quoted in a 1994 *NewsBytes* article headlined "Info Highway Should Meet Multiple 'Visions'"

To many people, some of what occurs will seem wasteful, disgusting, obscene, sexist, racist, even criminal. . . . The possibility remains that the NII could turn into a largely one-way street, one where "consumers" receive information but will not have freedom to retransmit or alter it. This is the "500 channels of TV" model, the worst scenario for the future because it implies an audience composed of inert consumers and passive paracitizens, easily manipulated by any technically adept spin doctors.—Writer **Tom Maddox,** in his 1994 *Wilson Quarterly* article, "The Cultural Consequences of the Information Superhighway"

All the headlines about the digital, interactive, 500-channel, multi-megamedia blow-your-socks-off future are pure hype. Yes, all the wild Wall Street, through-the-roof, Crazy Eddie, cornucopia, shout-it-out-loud promo jobs are pure greed. It's all a joke. It's now official. I'm announcing the beginning of a convergence backlash. There will be no convergence. There will be no 500-channel future. There will be no $3 trillion mother of all industries. There will be no virtual sex. There will be no infobahn. None of it. . . . Why all the hype and the greed? Everyone wants to look smart. But no one knows what to do with these new technologies.—**Mark Stahlman,** president of New Media Associates, in his 1994 *Wired* article headlined "Backlash: The Infobahn Is a Big, Fat Joke"

The real value of an information network will not be in a 500-channel cable TV and other entertainment media for home users but will be driven by business. The network will help large work groups automate their organizations and use fiberoptics and other technologies to transmit video, audio, and data.—Sun Microsystems CEO **Scott McNealy,** quoted in a 1994 *Business Times of Singapore* article headlined "Future Is with Internet, Says Head of Sun"

Then there's the myth that our computer networks will bring diversity, culture, and novelty into our classrooms and homes. . . . They promise 500 channels that will let us pick an entertaining and informative program from hundreds of offerings. . . . A 500-channel cable system will surely deliver unfathomable and boundless mediocrity. And with more channels, production values will further decline, since there will be less money spent per program.—Physicist and Internet critic **Clifford Stoll,** in his 1995 book *Silicon Snake Oil*

The boundaries we have been using to define our allegiances—national, religious, cultural—will break down. The greatest enemy to any fundamentalist regime is media, because media acts like water, slowly eroding ideological barricades. Meanwhile, individualists fear the coming of a monoculture, as iconography from the West washes over the unique landscapes of particular regions, while integrationists fear that 500 separate channels of cable television and hundreds of thousands of Internet Newsgroups will break up our world into isolated segments of like-minded individuals. . . . Neither nightmare need occur. . . . I see human culture becoming like a biological culture . . . where many individuals link together for common purpose, and where this linking . . . augments each member's ability to influence the organism.

—Author **Douglas Rushkoff,** in a 1995 essay for *New Perspectives Quarterly* titled "Coral Reef Culture"

You may have 500 channels instead of 50, but you won't get 10 times as many people willing to pay a penny more for what they get over the air for free. . . . The future won't be 500 channels—it will be one channel, your channel. So instead of subscribing to some à la carte, 24-hour channel, you'll just get the show you want on demand, whenever you want it. It'll be W-cubed: whatever, wherever, and whenever I want it.—**Scott Sassa,** president of Turner Entertainment Group, quoted in a 1995 *Wired* article headlined "If Mass Media Is Obsolete, and Pointcasting the Future of Media, Why Would Anyone in Their Right Mind Want to Buy a Broadcast Network?"

Really what they're offering you is a mall. They want to give you an infomall where you pay for every bit of information you download, and you'll download from a menu that some corporation has assembled. It's like they talk in the States (U.S.) about the "500-channel universe," and how we're all are going to have so much cable, but what are they going to put on it? In Los Angeles you can have a hundred channels of cable on your television today, and you can flip through all of them, and there's no content! It's amazingly content-free. So I have great hopes for the Internet, very little hope for commercial versions, and I profoundly hope that the Internet will continue to be the basis of this sort of growth.—Author **William Gibson,** in a 1995 interview on the Swedish television program *Rapport*

Culturally, we may lose touch with the mainstream. There may be no more mainstream. Politically, we may find ourselves detached from the dialogue. Right now, the president can talk over all three networks to us at the same time. With 500 video channels, that won't be so easy.—Author **Les Brown,** quoted in a 1995 *Seattle Post-Intelligencer* article headlined "Scholars Try to Measure the Impact"

When people talk about 500-channel TV, they mean 500 parallel streams. They don't mean one program after another, broadcast in one five-hundredth of real time. You don't download TV; you join an ongoing program. . . . When you buy a can of Coke, you are paying a few cents for the drink and the can, and nanodollars for television advertising. No doubt, the means of financing the bits will look strange to our great-great grandchildren. But for today, it's what makes television work. Eventually, we'll find new economic models, probably based on advertising and transactions. Television will become more and more digital, no matter what. These are givens. So it makes no sense to think of the TV and the PC as anything but one and the same. It's time TV manufacturers invested in the future, not the past—by making PCs, not TVs.—MIT Media Lab co-founder **Nicholas Negroponte**, in a 1995 *Wired* column headlined "Bit by Bit, PCs Are Becoming TVs. Or Is it the Other Way Around?"

INFORMATION GLUT IS NOT A STRONG ENOUGH DESCRIPTION

In the decades preceding the Internet boom of the 1990s, people were already concerned about the amounts of new information being created in what was becoming known as the "knowledge age" or the "information age." By the latter half of the 20th century, the capacity to create new information had surpassed humans' ability to take it all in. In 1966, Hubert Murray, a technical communication researcher, said at least 20 million words of technical information were being recorded in each 24-hour period. It would, he estimated, take a reader capable of reading 1,000 words per minute 1.5 months reading eight hours a day to get through one day of technical-writing output, and at the end of that stretch the reader would have fallen 5.5 years behind in keeping up with the output. This total only included a portion of the data being produced.

Richard Saul Wurman, an expert in the design and understanding of information, struck a chord with the public when he wrote about the worsening overload in *Information Anxiety*, a best-selling book in 1989. "Information anxiety is produced by the ever-widening gap between what we understand and what we think we should understand," he explained. "It is the black hole between data and knowledge, and it happens when information does not tell us what we want or need to know." To illustrate his point, he said a single copy of the November 13, 1987, edition of the *New York Times* totaled 1,612 pages, with more than 2 million lines and more than 12 million words. He pointed out that one weekday copy of the *Times* contained more information than the average 17th-century Englishman was likely to encounter in a lifetime.

A group of books on the infoglut topic followed in the 1990s, including titles such as *Data Smog* and *Survive Information Overload*. Wurman, who had originated the phrase "information architecture" in 1976, was attuned to his times, and rightfully so; since 1984 he had directed the TED conferences—which are still bringing together experts and entrepreneurs in technology (T), entertainment (E), and design (D). In the early 1990s, these by-invitation-only conferences were vital in building a synergy between some of the founding figures behind today's Internet. (It was at the 1992 TED conference that *Wired* co-founders Jane Metcalfe and Louis Rossetto gained MIT Media Lab founder Nicholas Negroponte's writing and backing for their magazine concept, selling him a 10 percent stake in exchange for $75,000 and his support.)

The information overload described by Wurman and others has become a way of life for most Americans, as they consume a daily avalanche of data through e-mail; IMs; text messages; phone calls; snail mail; print publications; television; and DVDs/videos.

Research firm Gartner estimated that by 2002 businesses around the world were spending as much as $30 billion annually on information-management systems. The Gartner researchers said that most corporate employees they interviewed told them the best-working information "filter" is a personal network of friends and colleagues who share information and offer tips, helping them cut through the clutter. Frank Odasz, director of Big Sky Telegraph, wrote about this phenomenon in a 1995 article titled "Issues in the Development of Community Cooperative Networks": "In the past, communities have formed to meet needs as a group that we cannot meet as individuals. Today, group protection from marauding animals or enemies is being replaced by group protection from the assaults of constant change and too much information. As information networking begins to enter mainstream society, each of us discovers new sources of data in our specific areas of interest, and suffers from the increasing pressure of information overload. . . . Networking increases the opportunities for collaboration, which helps protect individuals from information overload."

Market research firm IDC estimated that more than 60 billion e-mail messages would be sent annually by 2006. A good percentage of that would include information exchanges between friends or colleagues—a coping mechanism for dealing with information glut in order to avoid information anxiety.

A group of researchers at the University of California at Berkeley led by Peter Lyman and Hal R. Varian publishes a regularly updated report titled "How Much Information?" In the 2003 version, a study of the data from 2002, it was found that five exabytes of new print, film, magnetic, and optical storage media were created. Five exabytes, they explained, is the equivalent of 37,000 libraries the size of the Library of Congress, which they estimated to contain 136 terabytes of information (a terabyte is about the equivalent of 50,000 trees—each tree producing 80,500 sheets of paper). With the world population of the time at 6.3 billion, they said 800 megabytes of information per person had been produced, with 92 percent of it stored mostly on hard disks. According to Lyman and Varian, the staggering amount of worldwide information output had more than doubled since their 1999 survey, from two to five exabytes in just a few years.

Individuals today and in the future will find that a great deal of their success in life will be based on how well they learn to sift through information and extract what they need, sharing the important bits with others when they can. In a 1995 white paper for the U.S. Department of Education titled "The Evolution of Learning Devices," Chris Dede of George Mason University wrote: "The core skill needed in today's workplace is not foraging for data, but filtering a plethora of incoming information. The emerging literacy we all must master requires immersing ourselves in a sea of information and harvesting patterns of knowledge, just as fish extract oxygen from water via their gills. In this environment, educators must understand how to structure learning experiences that make this kind of immersion possible. Preparing students for full participation in 21st century society will require expanding the traditional definitions of literacy and rhetoric to encompass 'immersionlike' experiences of interacting with information."

INFORMATION OVERLOAD WEIGHS IN BY THE TON

We know you're busy, so we'll try to keep this section short. Following is a selection of predictions made between 1990 and 1995 regarding the overwhelming amount of information available in the age of networked communications. Could our gorging on an endless buffet of information cause everyone who heartily partakes to begin to suffer from a critical-thinking impediment?

What started out as a liberating stream has turned into a deluge of chaos. . . . Everything from telegraphy and photography in the 19th century to the silicon chip in the 20th has amplified the din of information, until matters have reached such proportions today that for the average person, information no longer has any relation to the solution of problems. . . . Our defenses against information glut have broken down; our information immune system is inoperable. We don't know how to filter it out; we don't know how to reduce it; we don't know to use it.—**Neil Postman,** in his 1990 speech "Informing Ourselves to Death"

A cloud . . . is building, a hurricane of over-information that threatens, if it continues, to serve as nothing more than nonsense deforesting large tracts of our national acreage. . . . To focus the amount of information we are producing on a weekly basis, which probably exceeds that produced in most of the preceding centuries, would take an enormous lens, or perhaps a million rather tiny ones.—*Boardwatch* editor **Jack Rickard,** in a 1991 panel presentation at the Conference on Computers, Freedom, and Privacy

One detects amid the hurly-burly of contemporary life a new constellation of feelings or sensibilities, a new pattern of self-consciousness. . . . Through an array of newly emerging technologies the world of relationships becomes increasingly saturated. We engage in greater numbers of relationships, in a greater variety of forms, and with greater intensities than ever before. With the multiplication of relationships also comes a transformation in the social capacities of the individual—both in knowing how and knowing that. . . . A multiphrenic condition emerges in which one swims in ever-shifting, concatenating, and contentious current of being. One bears the burden of an increasing array of thoughts, of self-doubts and irrationalities. The possibility for committed romanticism or strong and single-minded modernism recedes, and the way is opened for the postmodern being.—**Kenneth J. Gergen,** in his 1991 book *The Saturated Self,* which is excerpted in the 1997 book *Computers, Ethics, and Society*

Our saturation with information we don't need, with noise that increasingly rattles us, will begin to pall. . . . We will come ever more to resent the time it steals from our lives. Very few people any more want a real superhighway through their neighborhood.—Author **Bill McKibben,** quoted in a 1993 *New York Times* article headlined "We Are the Wired"

We may end up suffering a little from information overload, or spend a little more time on the couch, but I see that as a symptom of our success. —Microsoft CEO **Bill Gates,** in a 1993 *Red Herring* magazine article headlined "An American Gladiator in a Digital Arena"

The promise of electronic access is plagued by the problems of privacy, data security, lack of availability of the needed hardware and software for the public user, cost, inconsistent data formats, a patchwork of pricing schemes, absence of metadata, and lack of expert support for the public user. But more fundamental than those problems is that government collects and saves far more information than even it needs, this information is often redundant, and it is of no interest to anyone.—Iowa state Sen. **Richard Varn**, in a 1993 *Information Today* article headlined "On the Way to a National Electronic Library"

Content volumes have always outstripped channel capacity. . . . The same will be true for digital networks. There will be no "dead air" on our new conduits. This doesn't mean that content is king though, for if history is any guide, most of it will be fungible commodity properties such as game shows or old movies today, or utter trash. . . . Now the "vast wasteland" . . . will be supplanted by a vaster wasteland brimming with utterly new forms of interactive cyberdreck. —Futurist **Paul Saffo**, in his 1994 *Wired* magazine essay "It's the Context, Stupid"

Computers multiply data; in fact, one study indicated that data would double 19 times between 1990 and 2000. How will anyone be able to find the information needed in this huge haystack? The world is already choking on data. . . . Where is the information we have lost in data?—Technology consultant **George Gilder**, in his 1994 book *Life After Television*

Most people generally make a false assumption that more bits are better. More is more. In truth, we want fewer bits, not more. . . . Just because bandwidth exists, don't squirt more bits at me. What I really need is intelligence in the network and in my receiver to filter and extract relevant information from a body of information that is orders of magnitude larger than anything I can digest. To achieve this we use a technique known as "interface agents." Imagine a future where your interface agent can read every newspaper and catch every broadcast on the planet, and then, from this, construct a personalized summary. Wouldn't that be more interesting than pumping more and more bits into your home?—MIT visionary **Nicholas Negroponte**, in a 1994 *Wired* column headlined "Less Is More: Interface Agents as Digital Butlers"

What happens if this little magazine-sized personal computer of mine holds a million documents? Once it is no longer a problem to see or read the documents, the problem becomes how do you find them? How do you organize them? That is when you need agents to act as facilitators, to go find things for you. You need ways for the system to alert you when information becomes available. I mean, half the stuff that's on any server you've never seen before. You don't even know what to look for because you don't know it exists, so you need a serendipitous way of finding things, of having things present themselves to you.—PDF inventor **John Warnock,** quoted in a 1994 *Wired* article titled "Scriptwriter: John Warnock, the Inventor of PostScript, Founder of Adobe Systems, Plots the Future of Media"

The fact is, the Internet is now large and rapidly outgrowing its quaint netiquette. The Internet is running up against new limits—not on computing power, not on bandwidth, but on the attention span of its citizens.—Ethernet inventor **Bob Metcalfe,** in a 1994 *InfoWorld* article headlined "Advertising Can Save the Internet from Becoming a Utopia Gone Sour"

There's a kind of gymnastic ability just mentally that's needed to stay [up to speed with the changes new technologies, especially the Internet are bringing], so there is a frightening aspect to it. I think part of what we'll be doing over the next decade or so is figuring out some ways to somehow make ourselves comfortable with that, maybe by slowing down the pace of change.—Internet pioneer **Stewart Brand,** in a 1995 PBS-TV interview program titled *High Stakes in Cyberspace*

The Internet could literally be buried in a flurry of electronic junk mail. —Electronic Privacy Information Center founder **Marc Rotenberg,** quoted in a 1995 *Business Week* article titled "Law and Order in Cyberspace"

[Ten years from now the Internet is] going to be whatever big research system is possible. . . . My belief is that it will be a billion-dollar business in the early 21st century. And what is it? It's not Web fetching, which is just straight access, it's not library search, which is just what you're going to see in the next years when you can put up a big collection and actually search it. It's going to be correlation, analysis, coming in with a real problem and being able to look through many, many different sources and say, this thing here and this thing here combined in this certain way solves my problem. So we're going to talk about cross-correlation, generic community systems and spaces not networks.

—Bruce Schatz, principal investigator in the Digital Libraries project at the University of Illinois at Urbana-Champaign, in a 1995 keynote lecture for the American Society for Information Science

As information becomes cheap, learning what not to read or connect to becomes more valuable.... The technologies of disconnection [search tools, filters, and digesters that refine and manage data become vital].... And more people will subscribe to small information salons rather than cosmopolitan services like CompuServe.—*Wired* editor **Kevin Kelly,** quoted in a 1995 *San Francisco Chronicle* article headlined "Perils Await the Unwary on the Cyber-Frontier"

Without gatekeepers, without editors, the information highway has become a frustrating traffic jam. The Internet is akin to a noisy party line, cluttered with a cacophony of voices, many of them spewing endless reams of mindless, useless gibberish that ties up lines, slows down the system and makes finding a kernel of important discourse like locating that elusive needle in the haystack. Besides the garbage build-up, there are the oft-heard criticisms that the Internet has become the repository for pornography, hate literature, and even advice on how to build a bomb. Censorship of the 'Net has become a growing rallying cry.—Writer **Michele Mandel,** in a 1995 *Toronto Sun* article headlined "Net Takes Wrong Turn"

At parties, I'll scan the people: "not interesting, not interesting." Which is awful—sort of looking over their shoulders for the next person who might add value. It's a terrible, terrible thing to do.... There's a lack of depth and context and continuity in a lot of my face-to-face relationships. I think the whole quality of human interaction is changing.—Professional Internet searcher **Reva Basch,** in a 1995 *Wired* article headlined "Super Searcher"

We have access to mind-numbing amounts of data—the trash and the treasure, the ridiculous and the not-so-sublime.... We were promised that all these new options would enrich us. And yet even with this gluttony of choices, our diet is getting thinner.... The acceleration of daily life, this confusing mad rush to get ahead of the future, the speed of life in and about the media, is eroding our ability to gather the building blocks to do the real and necessary work.—Media executive **Barry Diller,** in a 1995 keynote speech for the American Magazine Conference

Globalization and the information revolution run the world. What can we do? Adapt or die. . . . We need to develop a higher tolerance for disorder to cope with this new world. That doesn't mean accepting disorder, but rather, working with the disorder to introduce order.—Intel CEO **Andy Grove,** in a 1995 *CandaOne* magazine article headlined "Case Example: Intel's Andy Grove"

In the next few decades, bits that describe other bits, tables of contents, indexes, and summaries will proliferate in digital broadcasting. . . . The result will be a bit stream with so much header information that your computer really can help you deal with the massive amounts of content. . . . The bits about the bits change broadcasting totally. They give you a handle by which to grab what interests you and provide the network with a means to ship them into any nook or cranny that wants them. The networks will finally learn what networking is about.—**Nicholas Negroponte,** in his 1995 book *Being Digital*

No amount of data, bandwidth, or processing power can substitute for inspired thought. . . . Today, those with the most information have the most power. This is patently false. . . . The Internet, that great digital Dumpster, confers not power, not prosperity, not perspicacity. Vice President Al Gore warned that we must not "divide our society into information haves and have nots." I'm not worried a bit; information is everywhere. You can take as much as you want.—Internet critic **Clifford Stoll,** in his 1995 book *Silicon Snake Oil*

[There is a] rising hierarchy with data at the bottom, information in the middle and knowledge at the top . . . [and] some would add intelligence or wisdom above that.—**David Ronfeldt,** of RAND, quoted in a 1995 *Government Technology* article headlined "Cyberspace 2020"

Technology may settle down fairly quickly because people just demand it; they can't stand all the change. Or it may settle down later because it evolves, it comes to a point where people are kind of comfortable and a number of things hold still and that's fine. Or it may simply not settle down at all. And then we're forced to just become used to always surfing a constant wave; it's never calm. And that would be a very interesting thing for civilization because we've never done that before. . . . The great thing about the future under these circumstances is that it is fundamentally unknowable, and that's both terrifying and very attractive.—**Stewart Brand,** in a 1995 PBS-TV interview *High Stakes on the Internet*

Voices of the Net

Zooming in on ten of the thousands of people who made a difference by addressing future concerns

How can you distinguish which voices rose in public discussions of the Internet in the early 1990s to become the most influential? You can't. It was the confluence of thousands of voices that gave the medium its form.

It is enlightening, however, to get to know a few of these key people a bit better, so here we zero in on 10 figures of the period between 1990 and 1995 who made a considerable mark on how the general public saw the Internet through their public statements about the revolutionary communications technology. These people had significant influence on the decisions of governmental and economic leaders of that time.

Included in the group are the two founders of the Electronic Frontier Foundation (John Perry Barlow and Mitchell Kapor); a trio of Internet illuminators (Howard Rheingold, Bruce Sterling, and Nicholas Negroponte)— some might call them gadflies—who crisscrossed the nation to give speeches and wrote reams of material about the networked world ahead; a researcher who was one of the few female voices prominent in press accounts at the time (Dorothy Denning); a technorealist who shared his concerns about the Internet (Clifford Stoll); a pair of forecasters/consultants who had firm views of what was to come (George Gilder and Paul Saffo); and a voice from the computer/networking industry (Gordon Bell). All of their quoted statements were made in the time span between 1990 and 1995.

JOHN PERRY BARLOW

John Perry Barlow helped found the Electronic Frontier Foundation (EFF) in 1990 with fellow WELL (Whole Earth 'Lectronic Link) member and Lotus Development Corporation founder Mitch Kapor to better respond in an organized manner to threats to free speech on the Internet. The WELL is a networked conferencing system—one of the early, well-established Internet communities—based in the San Francisco area, a U.S. region with a high concentration of technology industries. EFF is a nonprofit civil liberties organization that works to protect free expression and access to public online resources, and to promote responsibility in networked communications. The rapid engagement of EFF in national policy discussions immediately after its founding makes it one of the finest examples of grassroots political organizing on the Internet.

Barlow has described himself as a "cognitive dissident" and as a "free agent and peripheral visionary." Born to a prosperous ranching family near Jackson Hole, Wyoming, in 1947, he attended elementary school in a one-room schoolhouse. His 1969 bachelor's degree from Wesleyan University in Middletown, Connecticut, was in comparative religion. He owned and operated his family's cattle ranch in Wyoming for nearly two decades before selling it in 1988, and he collaborated in the song-writing process with rock group The Grateful Dead from the 1970s to 1995.

Barlow's 1990s writing on the future of the Internet—much of it reaching the mainstream public through *Wired* magazine—was thought-provoking and important. It made him a 1990s Internet "rock star" and enabled him to begin working as a consultant on the digital economy, copyright, and civil liberties. Barlow's article on the future of copyright, "The Economy of Ideas," is still a staple in law schools, and he has worked as a fellow at the Harvard Law School.

Following is a selection of statements made by Barlow about the future of the Internet.

> Just as limited bandwidth was the excuse for applying censorship to broadcast media, it appears that the zealous protection of intellectual property presents

the greatest threat to free digital expression.—1990, Electronic Frontier Foundation

The society we erect [in cyberspace] will probably be quite different from the one we now inhabit, given the fact that this one depends heavily on the physical property of things while the next one has no physical properties at all. Certain qualities should survive the transfer, however, and these include tolerance, respect for privacy of others, and a willingness to the treat one's fellows as something besides potential customers.—1991, *Communications of the ACM* (Association for Computing Machinery), in an article titled "Private Life in Cyberspace"

It is imaginable that, with the widespread use of digital cash and encrypted monetary exchange on the Global Net, economies the size of America's could appear as nothing but oceans of alphabet soup. Money laundering would no longer be necessary. The payment of taxes might become more or less voluntary.—1993, *Communications of the ACM*, "A Plain Text on Crypto Policy"

Humanity now seems bent on creating a world economy primarily based on goods that take no material form. In doing so, we may be eliminating any predictable connection between creators and a fair reward for the utility or pleasure others may find in their works. Without that connection, and without a fundamental change in consciousness to accommodate its loss, we are building our future on furor, litigation, and institutionalized evasion of payment except in response to raw force. We may return to the Bad Old Days of property. . . . We're going to have to look at information as though we'd never seen the stuff before. . . . The economy of the future will be based on relationship rather than possession. It will be continuous rather than sequential. And finally, in the years to come, most human exchange will be virtual rather than physical, consisting not of stuff but the stuff of which dreams are made. Our future business will be conducted in a world made more of verbs than nouns. —1994, *Wired* magazine, in an article titled "The Economy of Ideas"

Every time we make any sort of transaction in the digital environment, we smear our fingerprints all over Cyberspace. If we are to have any privacy in the future, we will need virtual "walls" made of cryptography.—1994, debate on AOL Forum

There's a great Bill Gibson line: "The future is already here, it's just unevenly distributed." There's tension from people who are on the (cyberspace) border. I'm afraid it will result in violence before it's all over. I want to see us thinking openly and seriously about how to avoid bloodshed. Because blood will be shed over this divide before it's over with. It's really just a question of how much. —1995, the *Rocky Mountain News*, Denver, in an article titled "Conflict Certain in Cyberspace"

In five years, everyone who is reading these words will have an e-mail address, other than the determined Luddites who also eschew the telephone and electricity. When we are all together in cyberspace we will see what the human spirit, and the basic desire to connect, can create there.—1995, the *London Guardian*, in an article titled "Howdy, Neighbours"

GORDON BELL

Gordon Bell proposed a plan for the online U.S. research and education network in a 1987 report to the Office of Science and Technology in response to a congressional request by Senator Al Gore. He established his impeccable credentials while working from 1960 to 1983 as a technology leader at Digital Equipment Corporation (where he led the development of the VAX computer and was vice president of research and development). He also helped Microsoft set up its first research laboratory in the early 1990s.

Bell was born in Kirksville, Missouri, in 1934, and earned bachelor's and master's degrees in electrical engineering from the Massachusetts Institute of Technology (MIT) in 1956 and 1957. While working in MIT's Engineering Speech Communications Laboratory in the late 1950s, he met MIT computer engineers Ken Olsen and Harlen Anderson. The three men started Digital Equipment Corporation, a leading developer of minicomputers and a key company in the early days of interactive computing.

Bell retired from DEC in 1983 and led the National Science Foundation's Information Superhighway initiative, working with the National Research Network panel that constructed the High-Performance Computer and Communications Initiative. He also helped found a number of small networked-computing start-up companies in the early 1990s.

As a consultant to Microsoft in the late 1990s and continuing into the new millennium, Bell helped develop the marriage of video and high-speed data networks and cultivated methods by which tech architecture could give desktop computers the same operating power as mainframes. He was one of the most savvy seers at predicting the growth numbers of Internet users and the

rise of the bandwidth offered by the Internet backbone over the past decade—
10 years after a speech he delivered at a 1995 InternetWorld conference, the
numbers he projected for those key growth measures have been proven to be
remarkably accurate.

Following is a selection of predictive statements made by Bell.

High-performance computers will enable wide use of digital images and image
processing, transforming a broad variety of disciplines, such as radiology, forecasting, urban geography and military intelligence. This paradigm shift will
transform every facet of science, engineering, and mathematics, starting with
the fundamental nature of education. Every home will have an unlimited laboratory in which to conduct experiments.—1992, *The Futurist*, in an article titled
"Computers in the 21st Century"

We can imagine a network with a range of PC-sized nodes costing between $500
and $5,000 that provides person-to-person communication, television and
when used together (including in parallel), an arbitrarily large computer. . . .
This architecture will drive out most other computer structures such as massively parallel computers, low-priced workstations and all but a few special-
purpose processors. This doomsday for hardware manufacturers will arrive before the next two generations of computer hardware play out at the end of the
decade. But it will be ideal for users.—1994, quoted by George Gilder in a *Forbes
ASAP* article titled "The Bandwidth Tidal Wave"

The sheer power of the microprocessor and increased magnetic capacity will
enable the PC to emerge as the most incredibly flexible product component civilization has ever had. PCs may be harnessed as ultra-powerful worldwide nodes
on the desktop, wristwatch-size personal digital assistants, voice-activated computers or television/computer combinations in the home.—1995, *Computerworld*, in an article titled "The View from Here"

Internet can be bigger than the PC and rival the telephone. . . . Internet 2.0 is
upsetting everyone's forecasts. . . . Next there will be lawsuits. Since we are all
spending hours browsing, there will be info-way addiction. And that's followed
by info-way regulations. . . . E-mail really is under-appreciated and under-
utilized and I think it'll be the dominant information carrier. It carries presentations, transactions, schedules, meetings and I hope bills. . . . We want a dial
tone for high-speed symmetrical links. Whatever the network's going to be, it
must be symmetrical. You can send to me, I can send to you at the same time

and data rate. That's an important part. Symmetry is the key. I want to allow bit warehouses or bit stores or bit places for audio, nice images, or television. And then, 4D so we can do virtual reality. That's a bandwidth question. With that we can do the tele stuff. That is, we can do the remote conferencing, remote work and remote business. I think those things are needed.—1995, InternetWorld keynote speech "It's Bandwith and Symmetry, Stupid!"

DOROTHY DENNING

Dorothy Denning was a professor and chair of computer science at Georgetown University in the 1990s, by which time she had been in the field of computer security and cryptography for two decades. Her research led her to become one of the most knowledgeable researchers of Internet fraud, hacking, espionage, and electronic sabotage. A 1990s profile in *Wired* magazine described her as the "scourge of wireheads."

Denning earned bachelor's and master's degrees in mathematics from the University of Michigan and a doctorate in computer science from Purdue University. Previous to her arrival at GU, she worked at Digital Equipment Corporation, SRI International, and Purdue. Her books include *Cryptography and Data Security* and *Information Warfare and Security*. She has written many Internet research studies. She was the first president of the International Association for Cryptologic Research. She testified before Congress on encryption policy and cyberterrorism and has worked in leadership positions on many networked computing panels. Her work illuminated the fact that organized crime and terrorist groups were using the Internet and encryption to communicate and share information.

As a leading proponent of the need for governments to be able to monitor criminal and terrorist activity on the Internet, she became a target of angry attacks leveled by foes of U.S. government attempts to regulate cryptography. Following a long career at Georgetown, she became a professor in the Department of Defense Analysis at the Naval Postgraduate School in Monterey, California.

Following is a selection of statements made by Denning about the future of the Internet.

Does the hacker ethic reflect a growing force in society that stands for greater sharing of resources and information—a reaffirmation of basic values in our constitution and laws? It is important that we examine the differences between the standards of hackers, systems managers, users, and the public. These differences may represent breakdowns in current practices, and may present new opportunities to design better policies and mechanisms for making computer resources and information more widely available.—1990, National Computer Security Conference, Washington, D.C.

In the context of the new milieu created by computers and networks, a new form of threat has emerged—the computer criminal capable of damaging or disrupting the electronic infrastructure, invading people's privacy, and performing industrial espionage. . . . A significant number of these hackers may go on to become serious computer criminals. To design an intervention that will discourage people from entering into criminal acts, we must first understand the hacker culture since it reveals the concerns of hackers that must be taken into account. We must also understand the concerns of companies and law enforcers. We must understand how all these perspectives interact. . . . Teaching computer ethics may help.—1991, *Communications of the ACM*, in an article titled "The United States vs. Craig Neidorf: A Viewpoint on Electronic Publishing, Constitutional Rights and Hacking"

Technology has been drifting in a direction that could shift the balance away from effective law enforcement and intelligence-gathering toward absolute individual privacy and corporate security. Since the consequences of doing so would pose a serious threat to society, I am not content to let this happen without serious consideration and public discussion. . . . The consequence of this choice will affect our personal safety, our right to live in a society where lawlessness is not tolerated, and the ability of law enforcement to prevent serious and often violent criminal activity.—1993, *Communications of the ACM*, in an article titled "Digital Communication Must Not Weaken Law Enforcement"

In the future, as the information superhighway looks more like an electronic marketplace, "digital cash" might be vulnerable to theft. "Burglars" might be able to break into a computer and download cash, and "muggers" might be able to rob intelligent agents that have been sent out on the network with cash to purchase information goods. . . . Encryption can protect against espionage, sabotage, and fraud. But it is a dual-edged sword in that it can also enable criminal activity and interfere with foreign intelligence operations. Thus, the role of encryption on the information superhighway poses a major dilemma. . . . E-mail fraud could become a serious problem as the information superhighway evolves into a major system of electronic commerce, with million-dollar contracts being negotiated and transacted through electronic mail. . . . If encryption comes into widespread use on the information superhighway, this could seriously jeopardize law enforcement and the public safety. Encryption is also a threat to foreign intelligence operations, and thus can affect national security.—1994, *Journal of Criminal Justice Education*, in an article titled "Crime and Crypto on the Information Superhighway"

It would be folly to let the capability to do electronic surveillance be completely overridden by technology. It's a much safer bet to put it into the system so that we can do it, to make sure that we have good procedural checks and laws to govern the use of that.—1995, CQ *Researcher*, in an article titled "Regulating the Internet"

GEORGE GILDER

George Gilder was a pioneer in the formulation of the theory of supply-side economics. In his major book *Microcosm* (1989), he explored the roots of the new electronic technologies. His book *Life After Television*, published by W. W. Norton (1992), is a prophecy of computers and telecommunications displacing the broadcast-TV empire. He followed it with another classic, *Telecosm*.

Gilder was born in New York City in 1939, and received his education at Exeter Academy and Harvard University, graduating in 1962. After studying with Henry Kissinger at Harvard, Gilder founded *Advance*, a journal of political thought based in Washington, D.C. He next served as a fellow at Harvard's Kennedy Institute of Politics and was editor of the *Ripon Forum*. During the 1960s, he wrote speeches for Nelson Rockefeller, George Romney, and Richard Nixon. In the 1970s, he studied the causes of poverty and wrote three books on the topic, including the bestseller *Wealth and Poverty* (1981).

Gilder pioneered the idea of supply-side economics in the 1980s, when he served as chairman of the Lehrman Institute's Economic Roundtable. He became a consultant for high-technology businesses and was quoted regularly by President Reagan. His focus in the 1990s was the future and impact of communications technologies, and he wrote, spoke, and served as a consultant to business and government bodies. He was also a contributing editor for *Forbes* magazine and a frequent writer for *The Economist, Harvard Business Review,* and the *Wall Street Journal*.

Following is a selection of predictive statements made by Gilder.

The telecomputer could revitalize public education by bringing the best teachers in the country to classrooms everywhere. More important, the telecomputer could encourage competition because it could make home schooling both feasible and attractive. To learn social skills, neighborhood children could gather in micro-schools run by parents, churches, or other local institutions. The competition of home schooling would either destroy the public school system or force it to become competitive with rival systems.—1991, quoted by **Roger Karraker** in an article in *The Whole Earth Review* titled "Highways of the Mind or Toll Roads Between Information Castles?"

The scarce resource is the human mind. People will be more valuable. People will get paid better. . . . It's utter garbage to say that our grandchildren won't live as well as we do. People who say this just don't see the technology. . . . There are all these wise-asses in Washington who really think that they can choose technologies. They think they know better. . . . It's always going to be that way. It's not going to change with Clinton and Gore. The dog technologies run to Washington, decked out like poodles. The politician is always the dog's best friend.— 1993, *Wired* magazine, in an article titled "When Bandwidth Is Free: The Dark Fiber Interview"

Computers will soon blow away the broadcast-television industry. But computers pose no such threat to newspapers. Indeed, the computer is the perfect complement to the newspaper. It enables the existing news industry to deliver its product in real time. It hugely increases the quantity of information that can be made available, including archives, maps, charts, and other supporting material. It opens the way to upgrading the news with full-screen photographs and videos. While hugely enhancing the richness and timeliness of the news, however, it empowers readers to use the "paper" the same way they do today—to browse and select stories and advertisements at their own time and pace. —1994, in the book *Life After Television*

Within the next five years, the entire American economy is going to be reshaped around these new digital networks. Telecommuting, teleconferencing, telemedicine, teleputing will change from buzzwords into basic fabric of business and life. . . . Within the next 10 years, this explosive technological advance in both networks and processors virtually guarantees that the personal-computer model of distributed intelligence and control will unseat the emperors of the mass media and blow away the television model of centralization. The teleputer—a revolutionary PC of the next decade—will give every household hacker the productive potential of a factory czar of the industrial era and the communications power of a broadcast tycoon of the television age. Broadcasting hierarchies will give way to computer heterarchies—peer networks in which the terminals are essentially equal in power and there is no center at all. . . . Families will regroup around the evolving silicon hearths of a new cottage economy.—1994, *National Review,* in an article titled "Net Gains: Information, Technology and Culture"

As the telephone network becomes a computer network, it will have to change, root and branch. All the assumptions of telephony will have to give way to radically different assumptions. Telephony will die. . . . TV ignores the reality that people are not inherently couch potatoes; given a chance, they talk back and interact. People have little in common except their prurient interests and morbid fears and anxieties. Necessarily aiming its fare at this lowest-common-denominator target, television gets worse and worse every year. Television is a tool of tyrants. Its overthrow will be a major force for freedom and individuality, culture and morality. That overthrow is at hand. . . . The force of microelectronics will blow apart all the monopolies, hierarchies, pyramids, and power grids of established industrial society. It will undermine all totalitarian regimes. Police states cannot endure under the advance of the computer because it increases the powers of the people far faster than the powers of surveillance. —1994, in the book *Life After Television*

The chief beneficiaries of all this invention . . . will be the people of the world, ascending to new pinnacles of prosperity in an Information Age. . . . Communications bandwidth is not only the secret of electronic progress. It is also the heart of economic growth, stretching the webs of interconnection that extend the reach of markets and the realms of opportunity. . . . The advance of the telecosm offers unprecedented hope to the masses of people whom the industrial revolution passed by.—1994, *Forbes ASAP,* in an article titled "The Bandwidth Tidal Wave"

The central event of the 20th century is the overthrow of matter. . . . The powers of mind are everywhere ascendant over the brute force of things. As humankind explores this new electronic frontier of knowledge, it must confront again the profound questions of how to organize itself for the common good. The meaning of freedom, structures of self-government, definition of property, nature of competition, conditions for cooperation, sense of community and nature of progress will each be redefined for the Knowledge Age—just as they were redefined for a new age of industry some 250 years ago. . . . Turning the economics of mass production inside out, new information technologies are driving the financial costs of diversity—both product and personal—down toward zero, "demassifying" our institutions and our culture. Accelerating demassification

creates the potential for vastly increased human freedom. It spells the death of the central institutional paradigm of modern life, the bureaucratic organization.
—1995, *Magna Carta for the Knowledge Age,* co-written with Esther Dyson, Jay Keyworth, and Alvin Toffler

MITCHELL KAPOR

Mitchell Kapor founded the Lotus Development Corporation and also founded the Electronic Frontier Foundation, a nonprofit Internet civil liberties organization, with WELL (Whole Earth 'Lectronic Link) member John Perry Barlow in 1990.

Born in Brooklyn, New York, in 1950, Kapor earned an interdisciplinary major in Cybernetics from Yale in 1971 by combining studies of psychology, linguistics, and computer science. He worked as a rock radio DJ, a transcendental meditation teacher, and an entry-level computer programmer before earning a master's degree in counseling in 1978. He was a mental health counselor for a short time, then attended the Sloan School of Management at MIT while also working as an independent software consultant. When he received a job offer from the publisher of VisiCalc, the world's first electronic spreadsheet, he left MIT one term short of graduation. He founded Lotus Development Corporation in 1982 with Jonathan Sachs, and they created one of the first "killer applications" of the personal computer, Lotus 1-2-3. By 1983, the company had revenues of $53 million, and by 1984, the revenues reached $156 million. By 1986, he'd successfully taken the company public, reaped his financial reward, and had left day-to-day management of Lotus. From 1987 through 1990, he was CEO of ON Technology, where he helped develop software for workgroups. He was also a founding investor of UUNET and Real Networks (the company behind the first popular streaming audio on the Internet), and has served on the board of directors for many Internet-oriented companies. Some of his most important work came as a co-founder of the Electronic Freedom Foundation.

Kapor served in vital roles in the unfolding of the Internet in the 1990s, chairing the Massachusetts Commission on Computer Technology and serving on the National Research Council's Computer Science and Technology Board and on the National Information Infrastructure Advisory Council. He was an adjunct professor at MIT's Media Lab from 1994 to 1996, teaching courses on democracy and the Internet and on digital communities. His writing and committee work during the decade of the 1990s helped formulate the United States' Internet policy.

Following is a selection of statements made by Kapor about the future of the Internet.

What is free speech, and what is merely data? What is a free press without paper and ink? What is a "place" in the world without tangible dimensions? How does one protect property which has no physical form and can be infinitely and easily reproduced? Can the history of one's personal business affairs properly belong to someone else? Can anyone morally claim to own knowledge itself? These are just a few of the questions for which neither law nor custom can provide concrete answers. In their absence, law-enforcement agencies like the Secret Service and FBI, acting at the disposal of large information corporations, are seeking to create legal precedents which would radically limit Constitutional application to digital media. [It] threatens to become a long, difficult, and philosophically obscure struggle between institutional control and individual liberty.—1990, Electronic Frontier Foundation founding statement, written with John Perry Barlow

At its best, the National Public Network would be the source of immense social benefits. As a means of increasing social cohesiveness, while retaining the diversity that is an American strength, the network could help revitalize this country's business and culture. . . . It will increase the amount of individual participation in common enterprise and politics. It could also galvanize a new set of relationships—business and personal—between Americans and the rest of the world. . . . [I recommend we] act now to create a level and competitive playing field for private network carriers, (whether for-profit or not-for-profit) to compete. Do not give a monopoly to any carrier. . . . Encourage information entrepreneurship through an open architecture (non-proprietary) platform, with low barriers to entry. . . . Everyone agrees in the abstract with universal service. . . . But that's only a platitude unless accompanied by an inclusive pricing plan. . . . The ideal means of accessing the NPN will not be a personal computer as we know it today, but a much simpler, streamlined information appliance— a hybrid of the telephone and the computer.—1991, Request for Comments #1259

If our legal and social institutions fail to adapt to new technology, basic access to the global electronic media could be seen as a privilege, granted to those who play by the strictest rules, rather than as a right held by anyone who needs to communicate. . . . Those parts of a system where damage would do the greatest harm—financial records, electronic mail, military data—should be protected. This involves installing more effective computer security measures, but it also means redefining the legal interpretations of copyright, intellectual property, computer crime, and privacy so that system users are protected against individual criminals and abuses by large institutions. These policies should balance the need for civil liberties against the need for a secure, orderly, protected electronic society.—1991, *Scientific American* article titled "Civil Liberties in Cyberspace: When Does Hacking Turn from an Exercise of Civil Liberties into Crime?"

Crucial doubts remain as to whether the re-wiring of America will result in Jeffersonian networks promoting the openness, freedom, and diversity that is the true promise of this technology. In the worst case, we could wind up with networks that have the principal effect of fostering addiction to a new generation of electronic narcotics. . . . Their principal themes revolving around instant gratification through sex, violence, or sexual violence; their uses and content determined by mega-corporations pushing mindless consumption of things we don't need and aren't good for us. . . . Purveyors of network services could simply decide that a business strategy that encourages the widest variety of content sources and originators will dramatically increase network usage. A few pennies per transaction will eventually add up to billions of dollars in revenue.—1993, *Wired* magazine, in an article titled "Where Is the Digital Highway Really Heading?"

The Internet today is still for computer weenies. But the problem will take care of itself. . . . Communities, whether virtual or physical, should be self-determining rather than determined by megacorporations.—1993, *The Nation,* in an article titled "The Whole World Is Talking"

NICHOLAS NEGROPONTE

Nicholas Negroponte, a co-founder of MIT's Media Lab and one of the foremost networked-communications promoters of his time, wrote one of the 1990s' bestselling books about the new future of communications: *Being Digital.* Negroponte had been an original investor who helped get *Wired* magazine off the ground, and his popular book evolved from the column he wrote for the back page of *Wired* magazine each month.

He was born on the Upper East Side of New York in 1943. He earned both his bachelor's and master's degrees in architecture at MIT. He joined the MIT faculty in 1966, and by 1968 he had founded MIT's Architecture Machine Group, a lab responsible for exploring new approaches to the human-computer interface. By the 1990s, he had helped found and serve as a major fundraiser for the building of MIT's Media Lab, which is focused on the development of future forms of human communication. The lab is supported by federal grants and monetary backing from more than 75 worldwide corporations—thus Negroponte's contemporaries sometimes called him the "P.T. Barnum of science." He also invested his own money in more than 40 technology start-up companies, and he served as a founding chairman of 2B1 Foundation, which is dedicated to supplying computer access to children in the poorest and most remote parts of the world.

Following is a selection of predictive statements made by Negroponte.

I do not believe there will be a Bit Police. The FCC is too smart. Its mandate is to see advanced information and entertainment-service proliferate in the public interest. There is simply no way to limit the freedom of bit radiation any more than the Romans could stop Christianity, even though a few brave and early data broadcasters will be eaten by the Washington lions in the process. —1993, *Wired* magazine, in a column titled "The Bit Police"

More bits per second is not an intrinsic good. In fact, more bandwidth can have the deleterious effect of swamping people and of allowing machines at the periphery to be dumb. . . . Fiber will come into being automatically through the forces of common sense and Mother Nature.—1993, *Wired* magazine, in a column titled "Debunking Bandwidth"

All of us are quite comfortable with the idea that an all-knowing agent might live in our television set, pocket, or automobile. We are rightly less sanguine about the possibility of such agents living in the greater network. All we need is a bunch of tattletale or culpable computer agents. Enough butlers and maids have testified against former employers for us to realize that our most trusted agents, by definition, know the most about us.—1994, *Wired* magazine, in a column titled "Less Is More: Interface Agents as Digital Butlers"

If the broadcast model is colliding with the Internet model, as I firmly believe it is, then each person can be an unlicensed TV station. . . . Most telecommunications executives understand the need for broadband into the home. (Recall, broadband, for me, is 1.5 to 6 Mbits per household member, not Gbits). What they cannot fathom is the need for a back channel of similar capacity.—1994, *Wired* magazine, in a column titled "Prime Time Is My Time"

Privacy may be more attainable in the world of bits than in the world of atoms. But we can also lose it faster if we don't pay attention. . . . The power of the word is extraordinary, and if the word is embodied as text, that, too, is powerful, regardless of whether the text lives as ink on pulp or signal on fiat-panel display. Words aren't going away, and I think the book/no-book argument is dumb once you realize that all we're talking about are variations in display technology. I'm not anti-book or anti-print; it's just that soon we're going to be doing our "printing" in a different medium.—1995, *Wired* magazine, in an article titled "Being Nicholas"

The Net makes it impossible to exercise scientific isolationism, even if governments want such a policy. We have no choice but to exercise the free trade of ideas. . . . For example, newly industrialized nations can no longer pretend they are too poor to reciprocate with basic, bold, and new ideas. . . . Now that ideas are shared almost instantly on the Net, it is even more important that Third World nations not be idea debtors—they should contribute to the scientific pool of human knowledge. . . . To think you have nothing to offer is to reject the coming idea economy. In the new balance of trade of ideas, very small players can contribute very big ideas.—1995, *Wired* magazine, in a column titled "The Balance of Trade of Ideas"

We are clueless about the ownership of bits. Copyright law will disintegrate. . . . Bits are bits indeed. But what they cost, who owns them, and how we interact with them are all up for grabs.—1995, *Wired* magazine, in a column titled "Being Digital: A Book (P)review"

My optimism comes from the empowering nature of being digital. The access, the mobility, and the ability to effect change are what will make the future so different from the present. The information superhighway may be mostly hype today, but it is an understatement about tomorrow. It will exist beyond people's wildest predictions. As children appropriate a global-information resource, and as they discover that only adults need learner's permits, we are bound to find new hope and dignity in places where very little existed before. . . . Being digital is different. We are not waiting on any invention. It is here. It is now.—1995, in his book *Being Digital*

HOWARD RHEINGOLD

Howard Rheingold was one of the first writers to illuminate the ideals and foibles of virtual communities. He wrote *Virtual Reality* (1991) and *The Virtual Community* (1993), books that were inspired by his exchanges on the WELL, one of the first computer-conference systems, located in the San Francisco area. He also was the editor of *Whole Earth Review* and the *Millennium Whole Earth Catalog*. He was hired in the mid-1990s to be the first editor of the online magazine *HotWired*—an offspring of *Wired* magazine—but quit quickly after a disagreement about the direction of the publication. In 1996, he founded and began publishing a webzine called *Electric Minds.*

Rheingold was born in Arizona in 1947 and earned a bachelor's degree from Reed College in 1968. He became an internationally recognized Internet personality in the 1990s, thanks to his writing and associated public appearances. His flamboyant style of dress in loud, colorful shirts accompanied by hand-painted shoes made him stand out. A continuing student of the sociology of technological change, he wrote *Smart Mobs: The Next Social Revolution* in 2002. Rheingold explains that "smart mobs" emerge when communication and computing technologies amplify human talents for cooperation, and he says the impacts of these uses are already both beneficial and destructive.

Following is a selection of 1990 to 1995 statements made by Rheingold about the future of the Internet.

The ability of groups of citizens to debate political issues is amplified enormously by instant, widespread access to facts that could support or refute assertions made in those debates. This kind of citizen-to-citizen discussion, backed up by facts available to all, could grow into the real basis for a possible electronic democracy of the future.—1991, from his book *The Virtual Community*

Will the future see an increasing gap between the information-rich and the information-poor? Access to the Net and access to college are going to be the gateways, everywhere, to a world of communications and information access far beyond what is accessible by traditional media. . . . It is the right of the citizens to remind elected policymakers that these technologies were created by people who believed that the power of computer technology can and should be made available to the entire population, not just to a priesthood. The future of the Net cannot be intelligently designed without paying attention to the intentions of those who originated it.—1991, from his book *The Virtual Community*

Laws that infringe equity of access to and freedom of expression in cyberspace could transform today's populist empowerment into yet another instrument of manipulation. Will "electronic democracy" be an accurate description of political empowerment that grows out of the screen of a computer? Or will it become a brilliant piece of disinfotainment, another means of manipulating emotions and manufacturing public opinion in the service of power? Human behavior in cyberspace, as we can observe it today on the Nets and in the BBSs, gives rise to important questions about the effects of communication technology on human values. What kinds of humans are we becoming in an increasingly

computer-mediated world, and do we have any control over that transformation? How have our definitions of "human" and "community" been under pressure to change to fit the specifications of a technology-guided civilization?
—1992, Electronic Frontier Foundation, from the article "A Slice of Life in My
Virtual Community"

Who controls what kinds of information is communicated in the international
networks where virtual communities live? Who censors, and what is censored?
Who safeguards the privacy of individuals in the face of technologies that make
it possible to amass and retrieve detailed personal information about every
member of a large population? The answers to these political questions might
make moot any more abstract questions about cultures in cyberspace.—1992,
Electronic Frontier Foundation

Every new communication technology—including the telephone—brings people together in new ways and distances them in others. If we are to make good
decisions as a society about a powerful new communication medium, we must
not fail to look at the human element.—1993, *Newsweek* magazine, in an article
titled "Cold Knowledge and Social Warmth"

Writers are going to work much more in teams in the future. . . . We haven't
reached the point where we've figured out how you can make a living at it [writ-
ing online]. But the power of the network is obvious. . . . I believe you can give
away a book online and make additional sales. . . . I can't imagine why anyone
would want to download 384 pages, print it out, bind it and carry it around,
much less try to sell it to someone.—1994, *Seattle Times,* in an article titled
"Communication Manifesto"

TCI could have tens of millions of television set-top boxes within a few years.
[Microsoft] Windows 95 is going to have the Microsoft Network bundled into
it, potentially reaching tens of millions of more users. Because of those huge
numbers, either company could wipe out Net culture and give itself tremendous
economic and cultural control. . . . [Biggest threat:] It's ripe for demagoguery
and attempts at censorship. That's the other side of our happy little utopian
dream. There are people who would destroy it, but so far they've either been
looking the other way or haven't succeeded yet in gaining leverage.—1995, *Inc.*
magazine, in an article titled "Net Alert"

PAUL SAFFO

In the 1990s, Paul Saffo was a director with the Institute for the Future, a
decades-old nonprofit forecasting foundation located in Menlo Park, Califor-

nia. The think tank was founded in 1968 by a group of former RAND Corporation researchers to perform consulting work for businesses and government entities, including telecommunications and consumer companies. Saffo said his work as a forecaster consists of helping people to understand what may happen and why, and he said the mission of IFTF is to "encourage stakeholders in society to think systematically about the long-range future."

Saffo attended Harvard and also earned a law degree—an LL.B.—from Cambridge University, and a degree in law from Stanford University. He worked for a Silicon Valley law firm for five years prior to beginning his career as a forecaster. His research regarding the long-term impact of new technologies on businesses and society made him an important voice of the "awe" stage of the Internet. In 1997, he was named as one of 100 "Global Leaders for Tomorrow" by the World Economic Forum. He served on the Stanford Law School Advisory Council on Science, Technology, and Society. His essays appeared in numerous major publications, including *Harvard Business Review,* the *Los Angeles Times,* the *New York Times,* and *Newsweek.* He also served as a member of the World Economic Forum Global Issues Group.

Following is a selection of predictive statements made by Saffo.

The personal computer may turn out to be like the horseless carriage, whose effect on our expectations was greater than its impact on our lives. PC users daydreamed of electronic cottages and global villages but settled for word processors and spreadsheets instead. . . . The next revolution will be shaped by new communications-rich workstations on our desktops and "information appliances"—inexpensive, portable high-performance information tools that will emerge from the collision of the computer and consumer electronics industries. Unlike PCs, these devices will be specialized tools for everyday tasks from calendar-keeping to communications. None will look remotely like a PC, and all will be cheap to the point of near-disposability. . . . The players most likely to shape this future will be the ones who shaped it a decade ago—small upstarts able to see the world in entirely new ways.—1991, *New York Times,* in an article titled "Farewell, PC—What's Next?"

Severed from unreliable paper, text has become all but inextinguishable. . . . paper may be on the skids, but text is eternal. Immortality may be the least of the surprises that this new medium of electronic text will deliver. Video enthusiasts are quick to argue that images are intrinsically more compelling than words, but they ignore a quality unique to text. While video is received by the eyes, text resonates in the mind. Text invites our minds to complete the word-based images it serves up, while video excludes such mental extensions. Until physical brain-to-machine links become a reality, text will offer the most direct of paths between the mind and the external world.—1993, *Wired* magazine, in an article titled "Hot New Medium: Text"

It is too early to tell what the digital counterculture will call itself, but the history of the hippies offers a clue. . . . the tekkies will arrive sometime in the mid-1990s, if not sooner. Watch the skies for a new comet—it will be digital, and its tail is likely to glow in Technicolor swirls. Its arrival will change our lives forever.—1993, *Wired* magazine, in an article titled "Cyberpunk R.I.P"

In the next decade, electronic mail is dead.—1994, *New York Times,* in an article titled "The Rise and Swift Fall of Cyber Literacy"

[An] avalanche of content . . . will make context the scarce resource. Consumers will pay serious money for anything that helps them sift and sort and gather the pearls that satisfy their fickle media hungers. The future belongs to neither the conduit or content players, but those who control the filtering, searching and sense-making tools.—1994, *Wired* magazine, in an article titled "It's the Context, Stupid"

BRUCE STERLING

Bruce Sterling, a nonfiction writer, consultant, and science fiction author based in Austin, Texas, wrote *Schismatrix* (1985), *Islands in the Net* (1988), and *The Difference Engine* (1990, co-authored with William Gibson), and edited *Mirrorshades: The Cyberpunk Anthology* (1986). His nonfiction *The Hacker Crackdown* (1992) is a book about computer crime and civil liberties that explains the events leading up to the founding of the Electronic Frontier Foundation. His writing for *Wired* magazine and his speechmaking at various Internet events and conferences of the early 1990s made him a key figure of the time.

He was born in Brownsville, Texas, in 1954, and attended the University of Texas at Austin from 1972 to 1976. In the 1990s, he wrote tech articles for *Fortune, Harper's, Details, Whole Earth Review,* and *Wired,* where he was a contributing writer since its founding. He published the nonfiction book *Tomorrow Now: Envisioning the Next Fifty Years* in 2002.

Following is a selection of statements made by Sterling about the future of the Internet.

[People] are very afraid of computer hackers, but I think mostly they are afraid of computers. . . . Computers are a challenge and a threat, and they're changing our society in ways that we can't control and don't understand. They're not to be trusted.—1991, *Compute!* magazine, in an article titled "Hack to the Future"

Too much access. By all means let's not provide our electronic networks with too much access. That might get dangerous. The networks might rot people's minds and corrupt their family values. . . . It's cultural struggle, political struggle, legal struggle. Extending the public right-to-know into cyberspace will be a mighty battle. It's an old war, a war librarians are used to, and I honor you for the free-expression battles you have won in the past. But the terrain of cyberspace is new terrain. I think that ground will have to be won all over again, megabyte by megabyte.—1992, speech at the Library and Information Technology Association Conference, Chicago

Computer networks worldwide will feature 3-D animated graphics, radio and cellular phone-links to portable computers, as well as fax, voice, and high-definition television. A multimedia global circus! Or so it's hoped—and planned. The real Internet of the future may bear very little resemblance to today's plans. Planning has never seemed to have much to do with the seething, fungal development of the Internet. After all, today's Internet bears little resemblance to those original grim plans for RAND's post-holocaust command grid. It's a fine and happy irony.—1993, *Magazine of Fantasy and Science Fiction,* in an article titled "Short History of the Internet"

A computer is something too close to the human brain for me to rest entirely content with someone patenting or copyrighting the process of its thought. . . . I don't think democracy will thrive in a milieu where vast empires of data are encrypted, restricted, proprietary, confidential, top-secret, and sensitive. I fear for the stability of a society that builds sand castles out of databits and tries to stop a real-world tide with royal commands. . . . I wish there were people in the Electronic Frontier whose moral integrity unquestionably matched the unleashed power of those digital machines. . . . The future is a dark road and our speed is headlong.—1994, Electronic Frontier Foundation reprint of an essay in *Science Fiction Eye* #10

Physically, you are very much here today and gone tomorrow, while cyberspace has become the anchor of your career and the linchpin of your reality. In 15 years, that will be a very common sentiment, so common that it will no longer seem odd or even remarkable.—1995, e-mail interview with Telecommunications International

Countries that have offshore money laundries are gonna have offshore data laundries. Countries that now have lousy oppressive governments and smart, determined terrorist revolutionaries are gonna have lousy oppressive governments and smart determined terrorist revolutionaries with computers. Not too long after that, they're going to have tyrannical revolutionary governments run by zealots with computers; then we're likely to see just how close to Big Brother a government can really get. Dealing with these people is going to be a big problem for us.—1995, *Wired* magazine, in an article titled "Good Cop, Bad Hacker"

Current trends in communications are leading toward a head-on collision between global networking and national governmental authority. At the moment a "twilight of sovereignty" scenario looks plausible and the situation definitely does not favor governments. Given current political instability worldwide, it's going to be a lot easier to make governments look like computer networks than it is to make the computer revolution the handmaiden of traditional governments. I make no judgment as to whether this is good or bad. After the revolution things will be different—not better, just different.—1995, e-mail interview with Telecommunications International

CLIFFORD STOLL

Clifford Stoll was an astrophysicist who also wrote the influential books *Silicon Snake Oil* (1995), *High Tech Heretic: Why Computers Don't Belong in the Classroom and Other Reflections by a Computer Contrarian* (1999), and *The Cuckoo's Egg*. An early network user because of his work as a researcher and scientist, Stoll made *Silicon Snake Oil* his platform for finding fault with the Internet hype of the early 1990s. He pointed out the pitfalls of a completely networked society and offered arguments in opposition to the hype.

Stoll earned his bachelor's degree in physics from the State University of New York at Buffalo, and his doctorate in astronomy was earned at the University of Arizona. While working as the systems manager at the Lawrence Berkeley Lab in the late 1980s, he discovered an error that he traced to the presence of a hacker on the system. He spent a year stalking the unauthorized user, uncovering a spy ring that had attempted to break into more than four hundred systems tied to national security, had succeeded in entering at least thirty of the systems, and had sold secrets to the KGB for cocaine and cash.

His first book, *The Cuckoo's Egg: Tracking a Spy Through the Maze of Computer Espionage*, is a retelling of that experience. Stoll made national headlines as the spy-catcher, and his well-known expertise in use of the Internet made his voice of concern all the stronger when he came out in the 1990s as an anti-Internet activist.

Following is a selection of predictive statements made by Stoll.

I see a wide gulf between the real networks that I use daily and the promised land of the information infrastructure. . . . Few aspects of daily life require computers, digital networks, or massive connectivity. They're irrelevant to cooking, driving, visiting, negotiating, eating, hiking, dancing and gossiping. You don't need a keyboard to bake bread, play touch football, piece a quilt, build a stone wall, recite a poem or say a prayer. . . . I'm saddened that so many accept the false promises of a hyper-hyped idea. Over-promoted, the small, intimate benefits of the Internet are being destroyed by their own success.—1995, in his book *Silicon Snake Oil*

Baloney. Do our computer pundits lack all common sense? The truth is no on-line database will replace your daily newspaper, no CD-ROM can take the place of a competent teacher and no computer network will change the way government works. . . . Bah. These expensive toys are difficult to use in classrooms and require extensive teacher training. Sure, kids love videogames—but think of your own experience: Can you recall even one educational filmstrip of decades past? I'll bet you remember the two or three great teachers who made a difference in your life.—1995, *Newsweek* magazine, in an article titled "The Internet? Bah!"

The heavily promoted information infrastructure addresses few social needs or business concerns. At the same time, it directly threatens precious parts of our society, including schools, libraries and social institutions. No birds sing. —1995, quoted in the *Buffalo News* in an article titled "Time to Exit the Information Superhighway?"

I doubt our offices will be replaced by minions working from home. The lack of meetings and personal interaction isolates workers and reduces loyalty. Nor is a house necessarily an efficient place to work, what with the constant interruptions and lack of office fixtures. Perhaps it'll work for jobs where one never has to meet anyone else, like data entry or telephone sales. What a way to turn a home into a prison.—1995, in his book *Silicon Snake Oil*

Local education budgets hardly cover salaries, books and paper. Computers address none of these problems. They're expensive, quickly become obsolete, and drain scarce capital budgets. Yet school administrators want them desperately.

What's wrong with this picture? . . . Today's standard connections, Ethernet and coaxial cable, will be obsolete within a decade.—1995, in his book *Silicon Snake Oil*

The falsehood of the Internet is that it will provide us with close, meaningful re-lationships, with cheap, good information and with useful life skills. Within each one of those promises is a grain of truth, but on balance they are simply false. . . . The Internet is not the key to the future. It's not going to provide great, wonderful information. Instead, it will continue to provide a rather mundane view of our very, very mundane world.—1995, *Maclean's* magazine, in an arti-cle titled "Reality Check"

11

The Threat to Freedom, to the Earth

As communications networks become all-seeing, some thinkers/theorists expect Big Brother or a robot takeover

Are science fiction writers clued in to the realities of the future? Most of them are futurists of a sort, and the body of their work views manmade machines being vital and often threatening presences at the social, political, and economic levels of our universe in the years to come. Today's technology elites join the great authors of fiction in viewing the future as one in which our machines may surpass our abilities to monitor them thoroughly, and some contemporary scientists go so far as to express the belief that this process will lead to human extinction or a massive war.

It could be said that our society's impressions of what a futuristic totalitarian government might look like were formed by George Orwell's classic book *Nineteen Eighty-Four,* and our vision of a future world in which robots are powerful and possibly threatening have been built out of the many science fiction works of Isaac Asimov.

BIG BROTHER SPEAKS TO THOSE CONCERNED
ABOUT PRIVACY AND THE INTERNET

The work of Eric Arthur Blair was referred to quite a bit in the early boom years of the Internet. The author of a brilliant futuristic satire written in the 1940s, Blair is better known as George Orwell. His book *Nineteen Eighty-Four* (published in 1949) is a key point of reference for any discussion of totalitarianism, dystopia, and government by surveillance. References to Orwell's fictional creations, including the Ministry of Truth, Big Brother, Newspeak, and ThoughtCrimes, have been peppered throughout hundreds of thousands of social commentaries, essays, speeches, and presentations since the book was published.

In his 1946 essay "Why I Write," Orwell explained that his experiences as a journalist reporting on the Spanish Civil War and World War II formed his concerns for the future of the world. "Every line of serious work that I have written since 1936 has been written, directly or indirectly, against totalitarianism and for democratic socialism, as I understand it," he explained.

Throughout the latter years of the 20th century, most of the people who wished to refer to a state in which individual freedoms are under threat by networked surveillance brought up Orwellian allegories. For instance, technology expert Paul Strassman echoed many stakeholders' concerns in his 1990 book *The Business Value of Computers*, writing,

> One should not view the electronic message that reveals, in a prescribed way, assumptions, links, and the structure of supporting nodes, as an unqualified gain. I can conceive of conditions where modifications of information can take place on a scale that does not approach even the wildest Orwellian visions. Network access to individual personal files makes it possible to reprocess and reedit collective memories, which will guarantee the removal of all nonconforming or undesirable facts. In its most objectionable form, metatext could reduce personal privacy through unwarranted intrusions into personal files.

When the Computer Professionals for Social Responsibility held its 1991 conference, "Computers, Freedom, and Privacy," the specter of the U.S. government as Orwell's Big Brother was a key motivation for the meeting. At one point during a panel session, *Boardwatch* editor Jack Rickard said,

> We have protections under the Constitution of the United States, and I want to see them applied online! If you're going to look at constraining Big Brother and

the technology he uses, you are always playing a catch-up game. Whatever you've just outlawed him using, there's something else coming down the pike. The approach to personal liberty should be the empowerment of the individual through their technology to be able to play the game as well. I don't think that simply constraining our government based on what happened last year is the most effective way to assure our personal freedoms. The new technologies offer you, as individuals, survival mechanisms to ensure your own liberty in the future.

Computer Professionals for Social Responsibility began in 1981 as a computer-message discussion group at the Xerox Palo Alto Research Center. The group gradually expanded, and it adopted its CPSR title in 1982. During the early 1980s, the group's primary concern was the military applications of computing technology. It opposed President Reagan's Strategic Defense Initiative, also known as the "Star Wars Defense System." It started The Privacy and Civil Liberties Project in 1986, opening a Washington office in 1988. In 1994, CPSR spun that office off into an independent organization titled the Electronic Privacy Information Center. Led by its director Marc Rotenberg, it has concentrated on the "Big Brother" issues surrounding digital networked communications. In the meantime, CPSR continues to center its advocacy work on all issues of social responsibility involving networked communications.

IT'S PRIVACY VERSUS LAW ENFORCEMENT
In a 1994 speech for law-enforcement officials at the High Technology Crime Investigation Association conference, technology writer Bruce Sterling posed the possibility that while networked communications can enhance some aspects of democracy, the technology can also be abused by criminals and tyrants. "Countries that have offshore money laundries are gonna have offshore data laundries," he said. "Countries that now have lousy oppressive governments and smart, determined terrorist revolutionaries are gonna have lousy oppressive governments and smart determined terrorist revolutionaries with computers. Not too long after that, they're going to have tyrannical revolutionary governments run by zealots with computers; then we're likely to see just how close to Big Brother a government can really get. Dealing with these people is going to be a big problem for us."

A lot of the "Big Brother" talk was and still is generated by the complexities involved in retaining such freedoms as personal privacy while also enabling law enforcement agents to help prevent threats to national security, computer viruses, and crime. During the early 1990s, the key controversy in this conflict was a Clinton administration plan to allow the government to gain access to encrypted, private digital information. The encryption device involved was called the Clipper Chip. David Farber, a high-speed networking expert, expressed his opposition in a 1994 online forum. "Clipper and the new digital telephony bills are a first step into what Orwell should have called 1994," he wrote during the digital discussion. "I see the slide ending with more and more government intervention in our private conversations. It will be the equivalent in cyberspace of having mikes in our living and bedrooms."

In a 1995 online paper debriefing his work on the White House Information Technology Task Force, federal technology administrator Randy Katz wrote about potential privacy problems (parenthetical references are Katz's):

> Despite all of the excitement associated with the InfoBahn, one potential "oil slick" is the government's encryption technology. A strong tension exists between the needs for personal privacy on the one hand and law enforcement on the other. It is possible for a senior administration official, Dr. Michael Nelson, special assistant within the White House Office of Science and Technology Policy, to utter, in the same speech (HPCC Symposium, March 1994) . . . the following: "We need to protect the American public (from terrorists, child pornographers, etc.)" . . . [and] "People should be able to use the NII to consult with their lawyer, their accountant or their doctor just as though they were in the same room." (Implying Big Brother should not be in that room.) While it has not been well publicized, a large number of system intrusions already involve cryptography. The intruders use it to cover their tracks. The encryption genie is already out of the bottle; a policy geared towards trying to control it appears to be unenforceable. The Administration has retreated from its initial proposal to use only National Security Agency algorithms and hardware. But the concept of key escrow remains a controversial cornerstone of the government's plan.

The Clipper Chip plan failed, under a barrage of criticism. The conflict between the right to privacy and the rights of law enforcement officials to seek information continues.

SOME SAY A LITTLE BIT OF BIG BROTHER IS A FAIR TRADEOFF

There were also 1990s voices that played down the warnings of an Orwellian nightmare ahead. Richard Sclove, author of the book *Democracy and Technology*, said in a 1992 *Christian Science Monitor* interview: "The technologies themselves offer ambiguous promises. It doesn't mean Big Brother has ar-rived. But it doesn't offer a democratic panacea either."

At a 1994 symposium titled "Free Speech and Privacy in the Information Age," Canadian writer Parker Barss Donham said:

> I remember a New Yorker cartoon from sometime in the 1960s or '70s, when computers were just getting a foothold in society. It showed a huge computer—an old Univac with whirring tape drives—sitting on a hilltop. A mob of farm-ers were rushing up the hill, attacking the computer with scythes and pitchforks. That's the image non-technical people had of computers, isn't it? Computers were feared, would be the Big Brother's tool, instruments of oppression, in-vaders of privacy, displacers of persons. All those dangers are still with us, and all those fears. The use of bar codes and scanners poses great challenges to a free society. But up until now, at least, computers have shown themselves to be ca-pable of empowering people in ways we did not foresee 20 years ago.

And in 1995, lawyer and technology consultant Peter Huber wrote in a *Columbia Journalism Review* article headlined "Big Brother, Goodbye": "Orwell's world, the world of computer and communications monopolies, will not be seen again in our lifetime. The loose ends and the forgotten comers have taken over. . . . The plugs and jacks and sockets have taken over the telescreen world; the Ministry is dead. Every untilled plug, every unconnected jack, is a loose end, a new entry into the network or an exit from it, a new soap box in Hyde Park, a new podium, a new microphone for poetry or prose, a new screen or telescreen for displaying private sentiment or fomenting sedition, for preach-ing the gospel, or peddling fresh bread."

Huber's point was proven at least twice by famous digital images released dur-ing the conflict in Iraq in 2004. Civilian contractor Tami Silicio took digital pic-tures of flag-draped coffins of fallen U.S. soldiers being returned home; the pictures, which violated a long-standing U.S. ban on such images, were published online and in newspapers and aired by the press. In addition, more than 1,800 images and video clips that illustrated abuses being committed at Iraq's Abu Ghraib prison brought to light the fact that U.S. military and civilian personnel were conspiring to torment prisoners. Some of the images were published on the Internet, the soldiers' actions were stopped, and they were sent to trial.

The images shared in these cases had a profound effect on various levels, and they proved a point: citizen monitors can now exert some degree of influence by playing Big Brother themselves. American comic Jon Stewart said on the television program *The Daily Show* that new technology now offers everyone "weapons of mass documentation" (a play on words tied to the search for "weapons of mass destruction" believed hidden in Iraq prior to the war).

Still, it's difficult to gauge the veracity of any image or sound clip in an age in which information made of bits can be assembled in innumerable ways; after London's *Daily Mirror* ran photos of British troops tormenting Iraqis in 2004, it was later revealed that they had been faked using photo-manipulation software—no such activities had actually taken place.

Following are seven additional revealing statements made between 1990 and 1995 in which references to Orwell's vision of unwavering surveillance comes into play.

It's hard to change information in books, but if we have everything online, then a somewhat untrustworthy group of people controlling the thing—which is what I think is what we have—gives us 1984.—**Alan Kay,** an originator of the graphical user interface, in a 1994 *Wired* article titled "Kay + Hillis"

I personally feel that the crusade for freedom and privacy in the digital age needs much better theories of the actual threats to freedom and privacy. Images like "Big Brother Is Watching You" really are not adequate, and better images of both the problems and the potential solutions will be a crucial part of the increasingly global campaign for democracy. . . . Somebody [must] throw some more light on the practices of the would-be monopolists, the companies whose business models are predicated on poorly regulated control of both carrier and content. This is not the free market in operation. Rather, it's large-scale "issues management" aimed at institutionalizing a set of anti-competitive regulatory structures. Issues management is the high-powered synthesis of lobbying, legal advocacy, public relations, and the quasi-intellectual work of "think tanks." . . . The cause of democracy would be greatly enhanced worldwide if the practices of issue management were thoroughly exposed and if clear, powerful metaphors for the process became as widespread as Big Brother and the Panopticon.—**Phil Agre,** editor of *The Network Observer,* in an introduction to the February 1994 issue

When every transaction leaves electronic footprints, pretty soon a computer knows things about us that we may want to keep hidden. . . . Out of this worry grows an absolutist position: don't let Big Brother collect information about me. . . . The Internet certainly will let different computers compare their data faster and will bring more databases online. Special programs can traverse the networks, ferreting out public information about any of us. A chilling discomfort. I'm not so worried. Today we're protected against this kind of database correlation by deep incompatibilities between different computer systems. . . . I suspect Big Brother won't have an easy time tracing us.—Clifford Stoll, in his 1995 book *Silicon Snake Oil*

If ubiquitous cameras tied into the information highway should prove to reduce serious crime dramatically in test communities, a real debate would begin over whether people fear surveillance more or less than they fear crime. It is difficult to imagine a government-sanctioned experiment along these lines in the United States because of the privacy issues it raises and the likelihood of constitutional challenges. However, opinion can change. It might take only a few more incidents like the bombing in Oklahoma City within the borders of the United States for attitudes toward strong privacy protection to shift. What today seems like digital Big Brother might one day become the norm if the alternative is being left to the mercy of terrorists and criminals.—Bill Gates, in his 1995 book *The Road Ahead*

Until very recently, George Orwell's nightmare tale of Big Brother . . . was the prevailing metaphor for the century to come. The frightening vision of Orwell's 1984 evaporated with the disintegration of the monolithic Nazi and Communist regimes. By contrast, the 21st century's defining image is more likely to have ordinary citizens using their personal telecommunications devices to keep Big Brother under continuing surveillance. . . . In the "smart" media world, information no longer flows only from one to many. Instead, it flows simultaneously and instantaneously in many directions, from the bottom up as well as from the top down.—Lawrence Grossman, in his 1995 book *The Electronic Republic*

With all of the Big Brother imagery, it is important to remember that these invasive devices also provide a service, one we often very much want. Many of us appreciate the convenience of credit cards, automatic teller machines, catalogue shopping, and cellular phones. Even the scary-sounding "smart roads" and "smart cards" may become things we cannot do without. . . . Having these

services and conveniences means that there will, as always, be a tradeoff. . . .
We will have all of the conveniences offered by computers, but we can never
again expect that our personal papers and communications can simply be
locked away from prying eyes and ears.—**Ellen Alderman and Caroline
Kennedy,** in their 1995 book *The Right to Privacy*

Here is the scenario—a conglomerate that includes a merger of Time-Warner,
QVC, Beatrice Foods, Microsoft, and Disney builds this new network—shin-
ing, gleaming, interactive, with fiber-optic cables running into every living
room—and for a reasonable fee they provide all the programs, all the channels, all the commercials, and allow us to shop from the convenience of our
BarcaLoungers; and they, of course, have access to demographic profiles of
our shopping preferences as well as our medical records, perhaps, and our educational records—certainly our preference in computer discussion groups. . . .
They will use this inside knowledge of our buying habits to drive us so deeply
into debt that our grandchildren will be born already owing this Time-
Warner-QVC-Microsoft-Disney aggregate every cent they will ever earn. Oh,
brother! Oh, Big Big Brother!—**Dinty Moore,** in the 1995 book *The Emperor's
New Clothes: The Naked Truth About the Internet Culture*

ROBOTS' ROLES IN THE FUTURE ARE A
CONCERN IN INTERNET'S EARLY BOOM

To some observers, another Big Brotherish specter was the idea of having "intelligent agents" set free on the network carrying people's private information. These agents were also called "knowbots" in the first stages of their
development at the Massachusetts Institute of Technology, and they cross
over, so to speak, combining both Orwell's Big Brother imagery and Isaac
Asimov's robots. The idea of creating artificial constructs that could do the
work of people goes back as far as ancient legends. The first recorded drawing of such a device has so far been attributed to Leonardo da Vinci, who included detailed drawings of a mechanical "knight" in his notebook late in the
15th century.

Construction of the first-known functioning robot is attributed to French
engineer Jacques de Vaucanson, who made a flute-playing figure in life-sized
form in 1738. Some say the first modern robot was a remotely operable boat
invented by Serb-American Nikola Tesla in 1898. Today limited-function industrial robots are used on assembly lines and in such applications as mining

and toxic-waste cleanup. Military applications include the "unmanned air vehicles" now being utilized as spy planes and bombers. Scientists have used rov-
ing robots to explore the surface of Mars. The field of social robots—the
prototypical domestic servant, for instance—is just beginning to make serious
progress. One example of such a device is Sony's Qrio, a bipedal humanoid capable of getting up after a fall.

ISAAC ASIMOV WROTE THE RULES FOR ROBOTS

The word *robot*—from the Czech noun meaning *labor*—was first used in a
1920 Czech play about chemically created workers. Isaac Asimov was the first
to use the term *robotics,* which he introduced in his 1942 short story
"Runaround." This was also the first time he referred to "the three laws of robotics." The Russian-born Asimov emigrated to the United States with his parents when he was 13. He wrote his first science fiction story in 1939, and
continued to create his popular fictional narratives as he earned his Ph.D. in
biochemistry. A prolific writer of fiction and nonfiction articles and books, he
is best known for his robot stories because they involved moral and ethical decisions. Many of these are collected in the 1950 book *I, Robot,* which was the
basis for a 2004 film with the same title.

In reality, significant progress in artificial-intelligence research is necessary

Prior to Asimov's work, most writers imagining a future with robots saw
them in black or white terms, as evil, corrupt, and dangerous or as benign and
under control. Asimov saw the grey areas and made them important in his
plotting. His "three laws," which most robots in his fiction followed, were: (1)
A robot may not harm a human being, or, through inaction, allow a human
being to come to harm. (2) A robot must obey the orders given to it by the human beings, except where such orders would conflict with the first law. (3) A
robot must protect its own existence, as long as such protection does not conflict with the first or second law.

In reality, significant progress in artificial-intelligence research is necessary
before any device can be programmed to follow Asimov's laws. Interestingly
enough, some of Asimov's later novels are plotted so harm is done *because* robots followed the laws to the letter, thus depriving humans of some needed
risk-taking and inventiveness.

WILL THE ROBOT AGENTS WHO DO OUR WORK BE NICE OR NASTY?

MIT Media Lab founder Nicholas Negroponte tried to assuage some of the
doubts about the future of robots entrusted with personal information in his

1995 book *Being Digital:* "A real estate agent builds a model of you from a succession of houses that fit your taste with varying degrees of success," he explained. "Now imagine a telephone-answering agent, a news agent, or an electronic-mail-managing agent. What they all have in common is the ability to model you. It is not just a matter of completing a questionnaire or having a fixed profile. Interface agents must learn and develop over time, like human friends and assistants. . . . A future interface agent is also seen as some centralized omniscient Orwellian character. A much more likely outcome is a collection of computer programs and personal appliances, each of which is pretty good at one thing and very good at communicating with the others."

William Mitchell of MIT devoted a chunk of his 1994 book *City of Bits* to a consideration of these network-information robots and Orwell's fictional future. "Orwell did not bother to think through the technical details, and [his] scheme would not really have worked—not with the primitive electronics that Orwell knew about, anyway," Mitchell wrote.

> Where would Big Brother have put all the corresponding monitors on the receiving end? Where would he have found the labor force to watch them all? How would he have sifted through and collated all that information? What actually happened was far more subtle and insidious. Instead of one Big Brother, we got a vast swarm of Little Brothers. Every computer input device became a potential recorder of our actions. Every digital transaction potentially left fingerprints somewhere in cyberspace. Huge databases of personal information began to accumulate. And the collation problem was solved; efficient software could be written to collect fragments of information from multiple locations in cyberspace and put them together to form remarkably complete pictures of how we were conducting our lives. We entered the era of dataveillance. . . . This is just the beginning; our lives have been leaving increasingly complete and detailed traces in cyberspace as two-way electronic communications devices have proliferated and diversified. . . . There is more of this to come.

In the 1993 article "The Machines Take Over," *Wired* writer Gareth Branwyn offers an excellent summation of Manuel de Landa's cautionary 1991 book *War in the Age of Intelligent Machines.* In the book, de Landa tracks the gradual transference of more and more of the U.S. military's centralized decision-making process to machines. "De Landa introduces us to a future robot historian, a second narrator of sorts that is charting its own genealogy," Branwyn writes.

He argues that computer hackers and visionary scientists have a prime opportunity to create "escape routes" that will thwart the efforts of the government and the military to "capture and enslave" liberatory technologies such as computer networks. . . . De Landa deftly applies some post-modern philosophy and the latest developments in the science of chaos and self-organizing systems. He looks at turbulent flows (migrations, crusades, invasions), singularities (points where order emerges from chaos), coherency, and many varieties of systemic noise, all as they apply to the emergence of military hardware, software, and wetware (the implant of technology directly into the body). . . . It is . . . a further exploration of the philosophy developed by Deleuze, Guattari, and Virilio.

IN MORAVEC'S FUTURE, HUMANS MAY
ALLOW ROBOTS TO MAKE THEM DISAPPEAR

The overlapping development of all types of networks built since the 1700s has complicated our world. Americans who lived in the 1700s may have had slow, simplistic methods for travel and communications, but they were fairly self-sufficient in comparison to Americans today. The transportation, energy, and communications improvements that have brought comfort, wealth, and prosperity over the decades have made people dependent upon burgeoning numbers of networks to accomplish their daily tasks—including an extensive power grid that makes it all work. There is no doubt that artificial intelligence will be taking on more and more duties as developments allow it to become an integral part of the networked world.

Professor Hans Moravec of the Carnegie Mellon Robotics Institute created a stir in the 1990s when he published his book *Mind Children: The Future of Robot and Human Intelligence.* In a 1995 interview with Charles Platt of *Wired* magazine titled "Superhumanism," Moravec explained his theory. He said that by 2030 robots will be able to learn and master skills and share the information with other robots throughout the network. Thus, the task of understanding the world will be shared among millions of robot minds, by 2040 robots will take over all work, and industry becomes hyperefficient. Humans don't have to work, they live in luxury, everything is really run by robots, and at this point, anything could happen.

"Now suppose a company goes out of business, leaving its [robotic] research division in space, where there's no supervision. The result is self-sustaining, superintelligent wildlife," Moravec told Platt. "You end up with [robots] forming a cyberspace where entities try to outsmart each other by causing their way of thinking to be more pervasive. Here's an ecology where all the dead-matter activity has been squeezed out and almost everything that happens is meaningful. You have this sphere of cyberspace with a robot shell, expanding outward toward Earth. . . . The competitive pressure toward miniaturization will result in activity on the subatomic level. They'll transform matter in some way; it will no longer be matter as we know it. . . . I don't think humanity will last long under these conditions."

But Moravec said there's a chance humankind could survive in some fashion. "Machine intelligences of the far future will . . . be able to reconstruct human society in every detail by tracing atomic events backward in time. . . . By this logic, our current 'reality' could be nothing more than a simulation produced by information entities. . . . The robots will re-create us any number of times, whereas the original version of our world exists, at most, only once. Therefore, statistically speaking, it's much more likely we're living in a vast simulation than in the original version."

Wired magazine carried a 1995 profile of Nicholas Negroponte titled "Being Nicholas" in which Negroponte told what he thought of Moravec's projected future. "Will we one day have robots running around who used to carry our groceries but are now hurling paving stones at us?" he asked. "I doubt it. I don't foresee a time when we are treated like pets by a culture of super computers that have us on invisible leashes. . . . Hans Moravec thinks that once computers are smarter than humans, we'll retire, and computers will become even smarter. I think the issue has more to do with consciousness and volition than being smart. Machines will be smarter than people, but I don't believe in artificial consciousness."

Even Moravec says everything might turn out all right. "We'll need police provisions so that legal companies will act to suppress rogues economically, or physically, if necessary," he said in the *Wired* interview. "And among the inprogrammed laws we'll need antitrust clauses to force dangerously large companies to divest into smaller entities. . . . Everything will depend on the way in which we create it. Crafting these machines and the corporate laws that control them is going to be the most important thing humanity ever does. You know, each age has an activity in which the best minds get involved. Crafting the laws, and their implementation, will be the thing to do in the 21st century."

COSMISTS AND TERRANS FACE OFF OVER ARTILECTS

Today, researchers and theorists working in the field of artificial intelligence call highly developed, independently operating devices "artilects." Professors and AI researchers Hugo de Garis and Kevin Warwick faced off on the issue at the 21st Alcatel Symposium in Zurich in March 2000.

Garis, an AI scientist who has worked in Belgium and Japan, is a Cosmist— a supporter of the building and proliferation of artilects. Warwick, a professor of cybernetics in the United Kingdom, is a Terran, a person who believes it is too dangerous to put such devices into play in the world. U.S. researchers Ray Kurzweil and Moravec articulated their Terran and Cosmist views in a similar debate in the United States just a week later. The scientists' intention in staging such events is to draw attention to the need for forethought in regard to the important moral and ethical issues involved in development of advanced artificial entities.

Bill Joy, chief science officer at Sun Microsystems, has cautioned that self-replicating nanodevices could eventually infect and destroy the technological infrastructure. And Garis told a *Wired* magazine interviewer in the 1997 article "The Architect of Man's Demise": "I see 21st century global politics dominated by the issue of species dominance."

Garis has also predicted that there may be a "gigadeath war—billions of dead" between Terrans, who fear to the destructive potential of artilect development, and Cosmists.

"I think humanity should build artilects—gods," Garis told *Wired*. "To choose never to do such a thing, would be a tragic mistake, I feel. We are too limited to be much as humans, but artilects have no limits. . . . I suspect that most of the advanced civilizations 'out there' have gone Cosmist, i.e. they have become artilects. Maybe there is a whole intergalactic community of artilects out there who don't bother communicating with biological forms because we are too primitive. Who knows?"

The following statements, made between 1990 and 1995 about networked communications, reveal the interest at that time in what was seen as the future of human-machine interaction.

FROM 1990 TO 1993—KAHN, RHEINGOLD, AND MORAVEC

A new class of sophisticated tools will have to be developed and used, such as knowledge robots, or "knowbots," as Dr. Kahn calls them. These otherworldly creatures will have ravenous appetites for information. They will be sent on fact-finding missions for humans, traveling at almost the speed of light to the appropriate electronic destinations and searching through appropriate databases. Prototype knowbots have already been created in software designed by scientists at Dr. Kahn's NRI [National Research Institute] and put to work, experimentally. In addition, refined computer hardware will have to be produced to handle the communications speed of gigabit networks and the volume of data that will accumulate as knowbots scan all the paper books and research journals on a virtually endless list of topics.—Reporter **John Markoff** paraphrases Internet pioneer Robert Kahn, in a 1990 *New York Times* article headlined "Creating a Giant Computer Highway"

Entire ecosystems of Net-spawned information-seeking robots will be circulating through the Net. These entities are formally more akin to robots (automata is the precise jargon) than to living creatures, but increasingly, automata are being designed to incorporate biological behaviors. The "worms" that can attack networks, and the "viruses" that pester computer users are the malevolent side of this trend. Knowbots and Rosebud are the benevolent side. In the environment of a heterogeneous, free-form Net, you are going to find both kinds. How you protect the community from dangers of attack without destroying the openness that makes the community valuable is a social problem.—Internet pioneer and social observer **Howard Rheingold**, in his 1991 book *The Virtual Community*

Because it will use resources more efficiently, a mature cyberspace of the distant future will be effectively much bigger than the present physical universe. While only an infinitesimal fraction of existing matter and space is doing interesting work; in a well-developed cyberspace every bit will be part of a relevant computation or will be storing a useful datum. Over time, more compact and faster ways of using space and matter will be invented and used to restructure the cyberspace, effectively increasing the amount of computational spacetime per unit of physical spacetime.—**Hans Moravec**, robotics expert, in a speech delivered at the 1992 Library and Information Technology Association Conference and reprinted in the book *Thinking Robots, An Aware Internet, and Cyberpunk Librarians*

FROM 1994—MITCHELL, MADDOX, AND GRUBER
Anticipate the moment at which all your personal electronic devices—headphone audio player, cellular telephone, pager, dictaphone, camcorder, personal digital assistant (PDA), electronic stylus, radiomodem, calculator, Loran positioning system, smart spectacles, VCR remote, data glove, electronic jog-

ging shoes that count your steps and flash warning signals at oncoming cars, medical monitoring system, pacemaker . . . and anything else that you might habitually wear or occasionally carry—can seamlessly be linked in a wireless bodynet that allows them to function as an integrated system and connects them to the worldwide digital network. . . . you will have acquired a collection of interchangeable, snap-in organs connected by exonerves . . . your nervous system will plug into the worldwide digital net. You will have become a modular, reconfigurable, infinitely extensible cyborg.—**William Mitchell**, in his 1994 book *City of Bits*

The next big jump in computing, potentially as important as the jump that created the programmable electronic computer, must be inspired by biology. . . . It may be that the successor to that type of machine is gestating far from the hotbeds of computerdom, in an obscure corner of the chemical business: a field called process control. . . . Such computers will not be much like human-programmed digital computers. . . . They will be less like the idiots that digital boxes are now, utterly dependent on flawless programming, and more like dogs: trainable, but with an inherent set of instincts and abilities, herding our processes and reactions and systems like a border collie runs a flock of sheep.—Technology writer **Michael Gruber**, in a 1994 *Wired* article headlined "Neurobotics: The Future of Computing"

Buildings themselves will become computers—the outcome of a long evolution. . . . They are getting electronic nervous systems—network connections, cabling in the woodwork, and information appliances. . . . computers will burst out of their boxes, walls will be wired, and the architectural works of the bitsphere will be less structures with chips than robots with foundations. . . . You will also begin to blend into the architecture. In other words, some of your electronic organs may be built into your surroundings. . . . "inhabitation" will take on a new meaning—one that has less to do with parking your bones in architecturally defined space and more with connecting your nervous system to nearby electronic organs. Your room and your home will become part of you, and you will become part of them. —**William Mitchell**, 1994, *City of Bits*

Building the NII, we create a vast and productive niche for the enlargement of [Manuel] de Landa's "machinic phylum," worlds in which machines can grow and evolve, and this eventually may have profound implications for human consciousness. Even in the relatively primitive forms it takes today, information technology seems to encourage a fixation on virtual rather than real experience—on technologically mediated perception, not direct apprehension. It can also saturate us in a hypnotic image-repertoire that works to render us passive and dream-struck no matter who, if anyone, controls it.—Writer **Tom Maddox,** in a 1994 *Wilson Quarterly* article titled "The Cultural Consequences of the Information Superhighway"

My software surrogates can . . . serve as my semiautonomous agents by tirelessly performing standard tasks. . . . A more maliciously conceived one might be programmed to roam the digital highways and byways, looking for trouble—for opportunities to corrupt the files of my enemies, to plunder valuable information, to eliminate rival agents, or to replicate itself endlessly and choke the system. Fritz Lang got it wrong: the robots in our future are not metallic Madonnas clanking around "Metropolis," but soft cyborgs slinking silently through the Net. The neuromans of William Gibson are a lot closer to the mark.—**William Mitchell,** 1994, *City of Bits*

FROM 1995—EPSTEIN, BERNERS-LEE, AND TED KACZYNSKI
A true AI will be a big, smart entity that will want to replicate itself and protect itself. It will mutate in some sense or other; copies will split off, and they'll replicate through the Net. . . . A virus is not a good analogy. Viruses are incredibly stupid. They're barely alive. A better analogy is an alien intelligence that lands here and tells us it's going to live with us, and we have to adjust. There'll be no way to turn them off, because they'll be moving through wires near the speed of light. What this means for the human species, I have no idea. I just know it has to happen.—**Robert Epstein,** founder of the Cambridge Center for Behavioral Studies, quoted in a 1995 *Wired* article titled "What's it Mean to Be Human, Anyway?"

In two years, today's WWW browsers will be as outdated as a lime-green polyester leisure suit. By the time this article is published, there should be a new crop of browsers that integrate the WWW, e-mail, netnews, remote login, and other network services. These browsers will accept powerful scripting languages from servers. This will enable browsers and users to interact in a variety of ways. Browsers will also interact with one another and with network services. Actually, the possibilities are limitless. The Web should also see a massive increase in the capability of WWW servers, and search engines should improve. Perhaps we will see the development of useful "knowbots"—the offspring of today's prototype robots, which merely build indices. These future knowbots could aggregate information, build summaries, and otherwise aggregate data and reduce it to information.—**Karl Auerbach and Chris Wellens,** in the 1995 *LAN* magazine article "Internet Evolution or Revolution?"

The truth is I haven't the faintest idea where it is going to be in five years' time. When the Web as an information space becomes an assumption, then it will be time for the next revolution. In five years' time the next revolution may have happened on top of the Web. It will happen within the Web. It may be mobile code. It may be robots working for you. It may be people finding ways of interacting politically.—**Tim Berners-Lee,** quoted in a 1995 *Computer Reseller News* article titled "Web Inventor Berners-Lee Speaks Out on Internet Future"

Welcome to the infinitesimal world of micromachines—MEMS, as they are known—where physical laws are turned upside down and an entire mechanism can fit easily on the head of a pin. . . . Not only are these "microlabs" smaller than anything that was imaginable a few years ago, but they can be fabricated 1,000 to the silicon wafer. . . . Some of the fantasies include free-ranging microrobots cruising through the bloodstream, reporting on conditions, and making repairs on a cellular level as well as microfactories creating entire tool chests of micromachines.—Writer **Richard Rappaport,** in his 1995 *Wired* article "Welcome to the Playground of Big Science"

The human race might easily permit itself to drift into a position of such dependence on the machines that it would have no practical choice but to accept all of the machines' decisions. As society and the problems that face it become more and more complex and machines become more and more intelligent, people will let machines make more of their decisions for them, simply because machine-made decisions will bring better results than man-made ones. Eventually a stage may be reached at which the decisions necessary to keep the system running will be so complex that human beings will be incapable of making them intelligently. At that stage the machines will be in effective control. People won't be able to just turn the machines off because they will be so dependent on them that turning them off would amount to suicide. On the other hand it is possible that human control over the machines may be retained.

In that case, the average man may have control over certain private machines of his own, such as his car or his personal computer, but control over large systems of machines will be in the hands of a tiny elite—just as it is today, but with two differences. Due to improved techniques the elite will have greater control over the masses; and because human work will no longer be necessary the masses will be superfluous, a useless burden on the system. If the elite is ruthless they may simply decide to exterminate the mass of humanity. If they are humane they may use propaganda or other psychological or biological techniques to reduce the birth rate until the mass of humanity becomes extinct, leaving the world to the elite. Or, if the elite consists of soft-hearted liberals, they may decide to play the role of good shepherds to the rest of the human race. They will see to it that everyone's physical needs are satisfied, that all children are raised under psychologically hygienic conditions, that everyone has a wholesome hobby to keep him busy, and that anyone who may become dissatisfied undergoes "treatment" to cure his "problem." Of course, life will be so purposeless that people will have to be biologically or psychologically engineered either to remove their need for the power process or to make them "sublimate" their drive for power into some harmless hobby. These engineered human beings may be happy in such a society, but they most certainly will not be free. They will have been reduced to the status of domestic animals. Needless to say, the scenarios outlined above do not exhaust all the possibilities. . . . It would be better to dump the whole stinking system and take the consequences.—**Theodore Kaczynski,** the man convicted of being the Unabomber, in a portion of his 1995 "manifesto"

12

The Future of Networks

The global mind doesn't need humans, but they may be able to use it if they'd like

Many theorists, philosophers, and scientists see the Internet as merely an early manifestation of what is to become what they describe as a collective consciousness, a neobiological civilization with a global mind—a godmind, or an unlocatable, omnipresent entity. Their concepts go far beyond the forms most world citizens would guess to be the future of artificial intelligence.

By the early 1990s, knowledgeable forecasters realized the Internet would grow to be more than an e-mail system and information-sharing device. They said Earth's complex systems were about to be exponentially raised to untraceable levels of interlocking, interweaving complexities.

Danny Hillis, an inventor of massively parallel computing, said in a 1995 interview with *Sun World* titled "Java Day: You Are There": "The Internet can be seen as an emergent organism. It wasn't engineered; it has grown. . . . Communication is taking place outside of the human mind. There is positive feedback from within the network. The technology starts to change as a result of its own processes. Communication takes place between computers that is meaningful to them. Just think, you'll be able to say to your grandchildren, 'I was there when all computers couldn't talk to each other.' But what is more likely: you'll be explaining your time to an applet that your grandchildren created to deal with their grandparents."

Simply described, a network is a collection of things that have a connection of some sort. Cutting-edge tech experts of the 1990s expected that in the future, networked intelligence would become so sophisticated it would begin to do most of our thinking for us. Information technology would become enveloped by biotechnology and nanotechnology, ushering in the use of self-replicating devices that are so small—made of thin films or extremely fine particles—they are at the invisible molecular or even subatomic size. The networked system would interweave and build upon itself.

These pervasive, powerful, intelligent devices could possibly be grown out of biological matter and would perform a multitude of sophisticated transactions at light speed. A self-organizing, autonomous tissue of networked intelligence would envelop the world. The sense of actively sitting and using a computer would disappear; human-computer interaction would be at the sensory level, built into our functions, built into our city structures and tools, and seamlessly networked.

If this is true and we wish to successfully survive the transition to such a world, we'd better come to an understanding of the ways and means of networks as soon as possible.

It's important to pay attention to the work of biologists, psychologists, physicists, mathematicians, neuroscientists, engineers, and social scientists who have enlarged the study of what has been called "Gaia theory," "the theory of complexity," "dynamical systems theory," "network dynamics," or "the web of life."

DAWKINS, MEMES, AND DIGITAL ECOSYSTEMS

British zoologist Richard Dawkins argued in his 1976 book *The Selfish Gene* that the world's networks follow a code, replicating and effecting change in achieving their evolution and survival. He proposed the evolutionary transfer of culture in his theory. His work inspired scientists such as Hillis and others who hope to create digital ecosystems in which software is spawned. Dawkins originated the term *meme* (rhymes with seem) to describe a self-propagating unit of cultural evolution. Analogous to the gene, it is a unit of information that is passed along to the next generation. Dawkins said cultures evolve in the same way that physical constructs do. And just as genes may mutate over time, memes can undergo alterations. Today's science of memetics applies the concepts of population genetics to a group's culture. It is one of the ways in which networks are now being studied.

After Nobel-winning physicist Leon Lederman half-jested that the goal of physics was to come up with an abbreviated equation to completely describe the universe—compact enough to fit on a T-shirt—Dawkins rose to the challenge and offered the now famous banner: "Life results from the non-random survival of randomly varying replicators." One of Dawkins's key arguments is that the primary factor in the evolution of a construct is not the gene or the meme, but rather it is the replicator. Dawkins's work as a zoologist inspired scientists in other disciplines to recognize digital technology's potential as a medium for the planet's biotechnological evolution.

After interviewing Dawkins for a 1995 article for *Wired* magazine, writer Michael Schrage paraphrased him, writing, "All life, at its core, is a process of digital-information transfer. . . . The rise of cheap processors and parallel architectures creates the ideal digital ecosystems to spawn software rather than build it. Nature—not rational cognitive planning—becomes the guiding force for the next generation of software solutions."

THE SMALL-WORLD PHENOMENON AND SIX DEGREES OF KEVIN BACON

Networks can be found in space, deep underground, and in our oceans; they can be found in the inner workings of the systems of every living organism; they can be found in our highways, shipping lanes, power grid, and communications systems; they can be found in social, political, and economic systems— in myriad relationships between living things, from bees in a hive to bikers in the Tour de France to members of a terror organization.

Students of social networks—that includes most of us, since we're all tied up personally in uncountable such multilayered networks—have probably heard of "Six Degrees of Kevin Bacon." When people play this amusing game, they try to show off their knowledge of pop culture by explaining the interweave of social connections between well-known figures. The game is based on the "small-world phenomenon"—this is the theory that every human is somehow socially networked with everyone else through a short chain of connections.

The first researcher to formally propose the theory was Ithiel de Sola Pool, in a work he completed in the 1950s with mathematician Manfred Kochen. The two men didn't have computers on which to run through models of their social-networks concept, but years later Harvard social psychologist Stanley

Milgram decided to develop it further. Milgram wound up publishing a study that popularized the idea of social networks in a May 1967 *Psychology Today* article.

Milgram's first research experiment wasn't well conceived, and its fascinating hypothesis led to only a few, limited replications of his study, but the data showed that 5.5 was the likely number of connections needed to tie one person to another in the United States. The number was rounded up to 6, since you can't connect with half of a person. John Guare read Milgram's work and wrote a 1991 play in which this scantily researched social networking theory took center stage; the play was a hit that later became a popular film and both were titled *Six Degrees of Separation.*

In the play, Guare's lead character Ouissa Kittredge explains what came to be the common public perception of the theory. "I read somewhere that everybody on this planet is separated by only six other people," she says. "Six degrees of separation between us and everyone else on this planet. The president of the United States, a gondolier in Venice—just fill in the names. I find that extremely comforting, that we're so close, but I also find it like Chinese water torture that we're so close, because you have to find the right six people to make the connection. It's not just big names—it's anyone. A native in a rain forest, a Tiero del Fuegan, an Eskimo. I am bound—you are bound—to everyone on this planet by a trail of six people. It's a profound thought."

The small-world phenomenon (as explained by Guare) inspired a popular party game, Six Degrees of Kevin Bacon, a brain-teasing exercise in which a player tries to associate other film stars with the B-list actor as quickly as possible in as few links as possible. Since Bacon has acted in more than 50 films, he can be connected to most modern movies.

Building off the ideas of Pool, Milgram, and even Guare, mathematicians Duncan Watts and Steven Strogatz built a mathematical model and proposed in 1998 that small-world networks exist throughout the universe—not simply in human relationships. As Watts explains in his 2003 book *Six Degrees*, using examples ranging from power grids to the neural networks of the earth-dwelling nematode *C. Elegans*, they proved that the addition of a small number of random links can transform a disconnected network, making it highly connected. They found that any network can be a small-world network "so long as it has some way of embodying order and yet retains a small amount of disorder."

SYNCHRONIZED CRICKETS LEAD TO NEW STUDIES OF NETWORKS

It all started out when Cornell University graduate student Watts began investigating the synchronization of chirping groups of snowy tree crickets, which can seem as if they are being guided to work in unison. "All of a sudden, the quaint urban myth (Six Degrees of Separation) that my father had related to me seemed terribly important," Watts wrote in *Six Degrees.* Watts and his adviser Strogatz developed a mathematical model to explain the small-world phenomenon. They tested it in many ways, including a mathematical look at Kevin Bacon and other actors. At the time, the total database of actors numbered 225,000. Watts and Strogatz found that every actor could be connected to every other actor in an average of less than four steps.

After Watts and Strogatz published their first paper on the work, Notre Dame physicist Albert-Laszlo Barabasi asked them to share their data sets. He and an assistant had decided to update Milgram's study, refine it, and enlarge it into a study of the Web. They found that any 1999 Web document was only 19 clicks away from every other document.

A few months later, Barabasi published a breakthrough paper in the journal *Science* that answered more "small world" questions. In it, Barabasi coined the term *scale-free* to describe a network in which connectivity is uneven. Most scale-free networks are large. The Internet is an example, and so are the world's power grid and sexually transmitted diseases. These networks have a few highly connected "super nodes" or "hubs," but most nodes are weakly connected. For instance, new pages are being created all the time on the World Wide Web, and as they are created most of them will likely provide links to other popular pages—hubs. The less popular pages are weakly connected; the popular pages are hubs. This phenomenon can be seen in human relationships—popular people who connect in a number of cultural groups are hubs. There are many scale-free networks in the world.

Barabasi explains in his 2002 book, *Linked,* that the six- and 19-degrees answers mislead people into thinking that things are easy to find in a small world. He said,

> Not only is the desired person or document six/19 links away, but so are all the
> people or documents. . . . Since the average number of links on any given Web
> document was around seven, this means that while we can follow only seven
> links from the first page, there are 49 documents two clicks away, 343 three

clicks away, and so on. By the time we reach the nodes that are exactly 19 degrees away, in principle we would have checked 10 to the 16th power documents, 10 million times more than the total number of pages on the Web (at that time). The contradiction has an easy resolution: Some of the links we meet along the road will point back to pages we have seen before. But even if it only takes one second to check a document it would still take over 300 million years to get to all the documents that are 19 clicks away! [Which also reinforces Wurman's concern over information overload.]

Watts, Barabasi, and many other scientists from various disciplines have now come together to study the similarities and differences of networks of every type and size. "Networks share resources and distribute loads, and they also spread disease and transmit failure," Watts wrote in *Six Degrees.* "By specifying precisely how connected systems are connected, and by drawing explicit relationships between the structure of real networks and the behavior of the systems they connect—like epidemics, fads, and organizational robustness—the science of networks can help us understand our world."

Where would network theory be today if Duncan Watts hadn't remembered chatting with his father about John Guare's play, and if he hadn't applied the idea of Six Degrees to the phenomenon of the synchronized chirping of crickets?

GAIA, FEYNMAN, AND MINIATURIZATION
It has long been recognized that the universe is made of self-organizing networks—sets of fluid, complex, and interdependent relationships nesting within other networks. Since the era of pre-Hellenic Greece, the word *gaia* has been used to refer to the planet Earth as being a single living system made up of innumerable complex networks. Both Pythagoras and Plato said, to know yourself is to know the universe.

Those who see Earth as Gaia say that just as your body is composed of billions of cells forming a single living being, so the uncountable molecules that make up all things on Earth work together as a living organism, an autopoietic network—a complex adaptive system that continually reproduces itself.

How will the human network help the planet adapt to the advances to come in material science; quantum computing, physics, information theory, and cosmology; genetic engineering; biotechnology; nanoelectronics; and molecular electronics? Could poet William Blake have had an idea of the future when he wrote in his "Auguries of Innocence" in 1802, "To see the world in a grain of sand, / and heaven in a wild flower, / To hold infinity in the palm of your hand, / and eternity in an hour"? The Romantics of Blake's time saw nature as "one great harmonious whole," as Goethe put it.

During the 20th century, the theory of complex networks was gradually built and advanced by many scientists from a variety of disciplines, including Russian Alexander Bogdanov, Austrian Ludwig von Bertalanffy, the original Gestalt psychologists such as Germans Max Wertheimer and Wolfgang Kohler, quantum physicists Werner Heisenberg and Henry Stapp, cyberneticists Norbert Weiner and Ross Ashby, Englishman James Lovelock, Chileans Francisco Varela and Humberto Maturana, and physicists Fritjof Capra *(The Web of Life)* and Mark Buchanan *(Nexus: Small Worlds and the Groundbreaking Science of Networks)*.

Weiner wrote in *The Human Use of Human Beings* in 1950, "We are but whirlpools in a river of ever-flowing water. We are not stuff that abides, but patterns that perpetuate themselves." Humanities professor Mark Taylor wrote in his 2001 book *The Moment of Complexity*: "Complex adaptive systems can help us to understand the interplay of self and world in contemporary network culture. . . . We are gradually discovering that we are, in effect, incarnations of worldwide webs and global networks."

Humankind has added technology to the networked systems that make patterns on Earth.

The future of communications networks is all about power, speed, and miniaturization. In 1959, Nobel-winning physicist Richard Feynman, a U.S. scientist known for his approachable, entertaining lecture style, delivered the first talk about what would come to be known as nanotechnology. It was titled "There's Plenty of Room at the Bottom." In it, he said, "What would the properties of materials be if we could really arrange the atoms the way we want them? They would be very interesting to investigate theoretically. I can't see exactly what would happen, but I can hardly doubt that when we have some *control* of the arrangement of things on a small scale we will get an enormously greater range of possible properties that substances can have, and of different things that we can do."

Nanotechnology has to do with technologies on a scale of one-millionth of a millimeter in size. K. Eric Drexler was the first to envision self-replicating nanobots in his 1986 book *Engines of Creation: The Coming Era of Nanotechnology*. He founded the first group—the Foresight Institute—aimed at helping society prepare for the changes expected to come due to the development of molecular nanotechnology.

In a 1993 article in *Wired* magazine titled "Wired Wonders," Drexler predicted, "All signs point to a revolution that advances to the limits set by natural law and the molecular graininess of matter. Trends in miniaturization point to remarkable results around 2015: Device sizes will shrink to molecular dimensions; switching energies will diminish to the scale of molecular vibrations. With devices like these, a million modern supercomputers could fit in your pocket. Detailed studies already show how such devices can work and how they can be made, using molecules as building blocks."

Rodney Brooks, the director of the Artificial Intelligence Laboratory at MIT, wrote a 2002 essay "The Merger of Flesh and Machines" for the book *The Next Fifty Years*, edited by John Brockman. In it, Brooks argued:

> The technology of our bodies and of our manufacturing will be generalized as the same thing. . . . Much of what we manufacture now will be grown in the future, through the use of genetically engineered organisms that carry out molecular manipulation under our digital control. Our bodies and the material in our factories will be the same. . . . there will be an alteration of our view of ourselves as a species; we will begin to see ourselves as simply part of the infrastructure of industry. While all the scientific and technical work proceeds, we will again and again be confronted with the same constellation of disturbing questions: What is it to be alive? What makes something "human"? What makes something "subhuman"? What is a superhuman? What changes can we accept in humanity?

The kind of high-powered, speedy, miniaturized technology Drexler and Brooks project is still on the horizon, but already the planet is being blanketed with layers of sensitive networked devices such as temperature gauges, traffic lights, cameras, and sensors in such a profusion that it is expected there will be 10,000 such machines per human before 2010. Will this eventually lead to a worldwide "skin" of biotechnology—a self-organizing, autonomous tissue of networked intelligence that could possibly become what some predictors call a "godmind"?

Stakeholders and skeptics had a great deal to say between 1990 and 1995 regarding the future possibilities as the network becomes molecularized and hyper-ubiquitous. Included below are a dozen more statements from Douglas Engelbart, Tim Berners-Lee, Sherry Turkle, David Porush, Robert Adrian, Christian Huitema, Kevin Kelly, and others.

If cyberspace is utopian it is because it opens the possibility of using the deterministic platform for unpredictable ends. . . . We might even grow a system large and complex and unstable enough to leap across that last possible bifurcation—autopoietically—into that strangest of all possible attractors, the godmind.—**David Porush,** co-director for an AI project at Rensselaer Polytechnic Institute in a 1992 Library and Information Technology Conference speech titled "Transcendence at the Interface: The Architecture of Cyborg Utopia—or—Cyberspace Utopoids as Postmodern Cargo Cult"

As we pursue significant capability improvement, we need to appreciate that we will be trying to affect the evolution of a very large and complex system that has a life and evolutionary dynamic of its own. Concurrent evolution of many parts of the system will be going on anyway (as it has for centuries). We will have to go along with that situation, and pursue our improvement objectives via facilitation and guidance of these evolutionary processes. Therefore, we should become especially oriented to pursuing improvement as a multi-element, co-evolution process. In particular, we need to give explicit attention to the co-evolution of the Tool System and the Human System.—Computing pioneer **Douglas Engelbart,** 1992, in an article titled "Toward High-Performance Organizations" on his Bootstrap Institute site

Does the logic of network existence entail radical schizophrenia—a shattering of the integral subject into an assemblage of aliases and agents? Could we hack immortality by storing our aliases and agents permanently on disk, to outlast our bodies? (William Gibson's cyberpunk antiheroes nonchalantly shuck their slow, obsolescent, high-maintenance meat machines as they port their psychic software to new generations of hardware.) Does resurrection reduce to restoration from backup?—**William Mitchell,** in his 1994 book *City of Bits*

As very large webs penetrate the made world, we see the first glimpses of what emerges from that Net—machines that become alive, smart, and evolve—a neo-biological civilization. There is a sense in which a global mind also emerges in a network culture. The global mind is the union of computer and nature—of telephones and human brains and more.—**Kevin Kelly,** in his 1994 book *Out of Control: The New Biology of Machines, Social Systems, and the Economic World*

Maybe there's some evolutionary force pushing us toward a complete exteriorization of our individual psychic landscapes, a mutual exposure. Clearly, we are wiring ourselves in, each to the other. We seem to be creating, through media and communications technology, what some have called a species-wide nervous system.—Mondo 2000 writers **R. U. Sirius** (real name, Ken Goffman) and **St. Jude** (real name, Judy Milhon), in a 1994 article they wrote for *Wired* titled "The Medium Is the Message, and the Message Is Voyeurism"

As a Net user, you start to view yourself as an individual cell in this growing global brain. Online users form connections to learn and grow, as brain cells do. Eventually we'll develop a global memory, a global ethic.—College professor **Tom Gentry**, quoted in a 1994 *Modesto Bee* article headlined "Info-Culture Technology"

There is a sense in which a global mind also emerges in a network culture . . . the union of computer and nature—of telephones and human brains and more. . . . The particular thoughts of the global mind—and its subsequent actions—will be out of our control and beyond our understanding. Thus network economics will breed a new spiritualism. Our primary difficulty in comprehending the global mind of a network culture will be that it does not have a central "I" to appeal to. No headquarters; no head. That will be most exasperating and discouraging. In the past, adventurous men have sought the Holy Grail, or the source of the Nile, or Prester John, or the secrets of the pyramids. In the future, the quest will be to find the "I am" of the global mind, the source of its coherence. Many souls will lose all they have searching for it—and many will be the theories of where the global mind's "I am" hides. But it will be a never-ending quest like the others before it.—**Kevin Kelly**, in his 1994 book *Out of Control: The New Biology of Machines, Social Systems, and the Economic World*

Even as people have come to greater acceptance of a kinship between computers and human minds, they have also begun to pursue a new set of boundary questions about things and people. After several decades of asking, "What does it mean to think?" the question at the end of the 20th century is, "What does it mean to be alive?" We are positioned for yet another romantic reaction, this time emphasizing biology, physical embodiment, the question of whether an artifact can be a life.—Researcher **Sherry Turkle**, in her 1995 book *Life on the Screen*

Suppose. . . . all these minor problems are cleared up, would we be seriously empowered as [Vannevar] Bush would like us to be, as a whole? Let's think about scaling problems. Let's think of some large numbers. The number of Web documents. The number of people in the world. The number of neurons in the brain. We're thinking of lots of things all connected together. Web objects, people and neurons all have the ability to have random associations. The neurons seem to work (on a good day) as an integrated team. The people do in parts. The Web documents just sit there. But pretty soon the Web documents will start getting up and wandering around. So when Web objects become mobile, and start wandering around and interacting with each other, would you now put much money on them making sense as a whole?—World Wide Web innovator **Tim Berners-Lee**, in a 1995 speech at MIT titled "Hypertext and Our Collective Destiny"

The network of networks could enable a million human brains to interconnect and form a sort of planetary superbrain. . . . By 2020, 12 billion machines from cardiac simulators to vehicle braking systems will be connected to the Internet.—Paraphrase of a statement made by **Christian Huitema,** a researcher with the Institut Francais de Recherche en Informatique et Automatismes, in a 1995 Tech Europe report from the Telecom T95 Conference in Geneva

One day the virtual world might win over the real world. These new technologies try to make virtual reality more powerful than actual reality, which is the true accident. The day when virtual reality becomes more powerful than reality will be the day of the big accident. Mankind never experienced such an extraordinary accident. . . . I must say that cyberspace is acting like God and deals with the idea of God who is, sees, and hears everything.—**Paul Virilio,** in a 1994 interview with *CTheory* titled "Cyberwar, God, and Television"

We also know that the computer networks control, with or without human presence, electric supplies, water supplies, transportation systems, inventories and accounting, telephone and communications networks, and the whole infrastructure of world finance—stock markets, insurance, and banking, not to mention government, corporate and military surveillance, and control programs. . . . The stupifying naivety of the technology-dazed but well-meaning, politically correct and liberal Internet user who believes that all problems will be solved when everyone is wired into the "World Wide Web" is symptomatic of the schizophrenia of (post-)modern media culture. . . . You are not in control of Cyberspace, it is not there for your comfort and convenience, and no

one is driving it. There is no suggestion in the notion of Cyberspace that, should human beings suddenly cease to exist—or destroy themselves in some nuclear folly—the network of machines that constitute Cyberspace would vanish with them. Cyberspace assumes that the machines we have built will soon, in some leap of almost magical synergy, break free of their creators to constitute, by means of the communications networks we are generously building for them, a universe or nature of an entirely new and different order.
—Canadian artist **Robert Adrian,** in a 1995 *Medien und Oeffentlichkeit* article titled "Infobahn Blues"

13

Nobody Knows You're a Dog

Or do they? Privacy issues on the Internet

"On the Internet, nobody knows you're a dog."

If you type the joke line of Peter Steiner's most-famous *New Yorker* cartoon into a search engine, you get thousands of potential links on which people have either reprinted the cartoon or made some sort of comment in relation to it.

After the cartoon was first published in 1993, the phrase became symbolic of the issues of personal identity and privacy on the Internet, and it will live forever as an online-culture touchstone because first it seemed to reflect the ability to be completely anonymous on the Internet, and only a few months later it became a tongue-in-cheek point of reference for people who wished to bemoan the fact that privacy is violated in the networked environment thanks to marketers' databases, cookies (site-visit tracking software), personalized spam (unwanted e-mail sales pitches and scams), and other evidence that anonymity seemed no longer possible. The cartoon's content is so general and it has been reprinted so often that people have come to assign varying meanings to its message.

Steiner's comic was the inspiration for an off-off Broadway play titled *On the Internet, Nobody Knows You're a Dog: Six People, Six Lies, One Internet*. It centered on a group of individuals who are unable to function IRL (In Real Life). The play was written in 1995 by Alan David Perkins, and its characters communicate by misrepresenting themselves in an online newsgroup, by e-mail, and through Internet Relay Chat. It has subsequently been staged in various locations all over the world.

THE SEQUEL ZEROES IN ON THE LACK OF PRIVACY ON THE INTERNET

By 2000, the "Nobody Knows" cartoon had gained so much notoriety *New York Times* reporter Glenn Fleishman wrote the feature "Cartoon Captures Spirit of the Internet" to explain it. "I feel a little like the person (whoever it is) who invented the smiley face," cartoonist Peter Steiner told Fleishman in an e-mail interview included in the article. "There wasn't any profound tapping into the zeitgeist. I guess, though, when you tap into the zeitgeist you don't necessarily know you're doing it."

A two-panel sequel to Steiner's cartoon by *Buffalo News* cartoonist Tom Toles appeared in 2000. It presented the updated view that there is no privacy on the Internet. In the first panel, two dogs are looking at a computer and one says, "The best thing about the Internet is they don't know you're a dog." In the second panel of the Toles cartoon, both dogs are looking at the computer screen, which reads, "You're a four-year-old German Shepherd-Schnauzer mix, likes to shop for rawhide chews, 213 visits to Lassie website, chatroom conversation 8-29-99 said third Lassie was the hottest, downloaded photos of third Lassie 10-12-99, e-mailed them to five other dogs whose identities are . . ."

Many of the people making comments about the future of the Internet in the early 1990s made references to the "Nobody Knows You're a Dog" cartoon.

In a 1994 article for *Red Herring* titled "New Media—What's Real and What's Not," Mark Stahlman, the president of New Media Associates, shared his take on the Internet and the future of new media, writing:

> Hurray for Hype! Hurray for Hoopla! What fun! Keep them confused. Keep them guessing. The biggest, most headlined, most ballyhooed techno-gold-rush in all of human history—the converged interactive digital info-highway—has endlessly entertained us all. . . . We all know the technologies are real. . . . Yet there's a nagging sense that the roadmap to the future has been too easily sketched, too hastily drawn. . . . The technologies are obvious; the businesses in which to use them are not. . . . The challenge is to invent a new industry. Who sells what to whom? What's the new business model? . . . What's different about the times ahead? What does it mean if on the Internet nobody knows you're a dog?

In the keynote speech at InternetWorld 1995, pioneering computer scientist Gordon Bell, formerly of Digital Equipment Corporation and at that time a research leader at Microsoft, told of his vision of the next version of the Internet—Internet-3—saying: "I don't think we'll have relative [fully functioning video] for awhile, but . . . they're going to know you're a dog if you're on Internet."

In a 1995 column for *Wired* magazine titled "Being Digital: A Book (P)review," Nicholas Negroponte, founder of MIT's Media Lab, wrote: "Being digital is positive. It can flatten organizations, globalize society, decentralize control, and help harmonize people in ways beyond not knowing whether you are a dog. In fact, there is a parallel . . . between open and closed systems and open and closed societies. In the same way that proprietary systems were the downfall of once great companies like Data General, Wang, and Prime, overly hierarchical and status-conscious societies will erode. The nation-state may go away. And the world benefits when people are able to compete with imagination rather than rank."

And in a fall 1995 column in *Educom Review,* John Gehl wrote:

> The Steiner cartoon is, of course, about the Internet, and the reason it's funny is that it uses silliness (talking dogs) to point out truthfully that the Internet is an

impersonal form of communication where "nobody knows you're a dog." So the cartoon is both funny and true. Just how funny is it, on a scale of 5? Pretty funny. Give it a 4. How true is it? Give it a 2. First of all, the Internet is no more anonymous than a telephone call, or a novel, or this column. Second, people do know when they run into a dog (or a fool) on the Internet, just as they do when they get a phone call from a fool, browse a novel by a fool, or read a column by —But if you don't mind I'd like to move on now to my next point. The larger falseness of the cartoon is its implicit suggestion that the Internet will automatically and effortlessly (electronically!) produce a new renaissance of public expression because the Simple Honest Internet will let us toss aside the Bad Old Publishers the way that Marxism did away with capitalism (though I must have been on vacation when that event took place).

PERSONAL COMMUNICATIONS ARE SUDDENLY LESS PRIVATE

In the 1990s, the operators of some websites began requiring people to register their names and personal information before being able to gain access; this became a point of argument in the early stages of the Internet privacy war. In his 2003 book about *Wired* magazine, *Wired: A Romance,* former Wired staffer Gary Wolf quotes a 1995 e-mail publisher Louis Rossetto sent out to employees of the magazine's spin-off website, *HotWired.* The e-mail was sent in defense of Rossetto's view that Hotwired should retain its user-registration requirement for readers of the online site. Some employees of Hotwired opposed it because it cut across the "traditional" values of a free and open Internet and seemed to contradict the *Wired* philosophy. Rossetto steadfastly defended the use of registration in personal conversations and in an e-mail missive in which he wrote: "I reject the idea that anonymity itself is actually an overriding benefit to the users. . . . Anonymity is not the good, privacy is. Confusing the two is ahistorical, knee-jerk ideology that is going to look damn foolish in very short order, along with a lot of other wishful thinking about the Net."

The very fact that Wolf saved this e-mail and was able to use it and other e-mails from his history at *Wired*—e-mails that were probably written by the senders with no idea that they would later be asked by Wolf if these missives could be used as content his book—shows how the bounds of personal privacy came under threat in new ways beginning in the 1990s. Since the earliest beginning of the written word, personal communications were considered to be private information, and these messages—usually written on paper—were not easily copied and shared. The advent of copy machines changed that minimally, but the digital Internet with its e-mail capabilities and nearly instant web-page updating shattered the concept of private communications. Anything you send to someone else in a digital form—such as an e-mail in which you ridicule your boss or in which you mention inside information about your business—can be instantly sent to thousands of people at no cost to the sender.

People who were born in the pre-Internet age were not socialized to be wary of such intrusions into what was once considered to be a personal space. As Ellen Alderman and Caroline Kennedy wrote in their 1995 book *The Right to Privacy*: "E-mail is not the U.S. Mail; a cellular phone is not a traditional phone; and our financial, medical, and other records are no longer stored in filing cabinets. If we want to secure our communications and personal profiles, we will have to become more vigilant. Even so, there will still be some intrusions beyond the control of contract, legislation, or technology. Then we will have to alter our expectations . . . we will ultimately have to change our idea of what we can reasonably expect to keep private."

PRIVACY INFRINGEMENTS TO FOLLOW YOU EVERYWHERE YOU GO

Readable digital implements known as ubiquitous-computing devices first came to the forefront of concern of digital-privacy advocates in the early 1990s. Scientists at Xerox's Palo Alto research facility were working on further development of what they called "active badges." These portable computing devices would allow a number of actions, including the tracking of people's whereabouts.

There was some concern about the privacy ramifications involved with being in constant touch with a home base that records your movements. Professor Stephen Doheny-Farina of Clarkson University wrote in a 1994 article for *Computer-Mediated Communication* titled "The Last Link": "Active badges should scare the daylights out of anyone. When it comes to connectivity, the employer must justify the surveillance. Everyone must assume that only extraordinary conditions merit surveillance. The requisite argument must not be, 'Why do you not want to wear the badge?' The requisite argument must be, 'Why do you want me to wear it?' We must demand that the burden of proof is on the watcher, not the watched."

Employers' use of such tracking devices to monitor employees' movements is seen as an obvious problem. "Fair Information Practices may be required to help ensure that active badges are a convenient technology which do not degrade people's working lives," wrote Rob Kling in the 1994 article "Fair Information Practices" for the journal *Computer-Mediated Communication*. "Other kinds of information practices, such as those in which location monitoring is non-reciprocal, and non-discretionary may help transform some workplaces into electronic cages." Roy Want, a scientist exploring active-badge technology at Xerox PARC, told a writer for *Wired* magazine in a 1995 article titled "You're Not Paranoid; They Really Are Watching You," "There are always these trade-offs between what's useful and what could be done to us. The benefits to be had are so great; we just have to be sure that the people who are in control respect our privacy."

Devices of a limited capability of this type are now known as RFID—radio-frequency identification devices—and they are proliferating at a staggering rate. These tiny computer chips include a miniature antenna that can send a radio signal. Some of them can output data, such as personal health information; others are made only to signal an identification number for tracking purposes—for instance, for airline baggage tracking.

These devices can now be produced cheaply and in such a way as to become invisible or nearly invisible parts of products you use every day. They are now commonly used to track such things as beer kegs, pallets, casino chips, and library books; they are used in auto-theft-prevention systems; and they are implanted in pets. California toll stations use RFID for toll collection. They can also be embedded under the skin; some humans who have severe health problems have had such devices implanted so they can retain a health record on their person at all times.

Commercial concerns find RFID useful in tracking supply-chain management and for inventory control, but privacy advocates see intrusions ahead in this technology. Activist websites have been established to warn people about the privacy issues tied to RFID, and researchers have been developing "blocker tags" that might disrupt the transmission of information from RFID tags.

The Electronic Frontier Foundation has published an online position paper on RFID. It reads, in part:

> Used improperly, RFID has the potential to jeopardize consumer privacy, reduce or eliminate purchasing anonymity, and threaten civil liberties. . . . RFID tags can be embedded into/onto objects and documents without the knowledge of the individual who obtains those items. As radio waves travel easily and silently through fabric, plastic, and other materials, it is possible to read RFID tags sewn into clothing or affixed to objects contained in purses, shopping bags, suitcases, and more. . . . Tags can be read from a distance, not restricted to line

of sight, by readers that can be incorporated invisibly into nearly any environment where human beings or items congregate. RFID readers have already been experimentally embedded into floor tiles, woven into carpeting and floor mats, hidden in doorways, and seamlessly incorporated into retail shelving and counters, making it virtually impossible for a consumer to know when or if he or she was being "scanned."

In 2003, the EFF requested that manufacturers and retailers agree to a "voluntary moratorium on the item-level RFID tagging of consumer items" until an assessment of the technology's effects could be studied. This request was not honored by industries involved in RFID.

It seems RFID in some form, with or without "blockers" or industry limitations of some sort, are going to continue to multiply. The next U.S. passport is likely to feature this sort of onboard memory chip and wireless antenna. It's expected to hold encrypted information, including your image, your name, and your birth date. Bruce Sterling warns about a future, souped-up RFID in a 2004 *Wired* magazine column titled "Dumbing Down Smart Objects": "Because it's tracked precisely in space and time, let's call it a 'spime.' . . . Get ready for: spime spam (vacuum cleaners that bellow ads for dust bags); spime-owner identity theft, fraud, malware, vandalism, and pranks; organized spime crime; software faults that make even a mop unusable; spime hazards . . . ; unpredictable emergent forms of networked spime behavior; objects that once were inert and are now expensive, fussy, fragile, hopelessly complex, and subversive of established values."

Sensors of many types—computer chips assigned to report data from various distances and generally designed to be invisible to the casual observer—have become ubiquitous. Since 1995, for instance, the San Diego freeway system's asphalt has carried sensors built to report traffic speed and densities for the California Department of Transportation. In 1997, Paul Saffo of The Institute of the Future wrote in his "Ten-Year Forecast":

What happens when we put eyes, ears, and sensory organs on devices? Inevitably, we are going to ask those devices to respond to what they "see," to manipulate the world around them. . . . This has profound implications. Two parallel universes exist today—the everyday analog universe we inhabit, and a newer digital universe created by humans, but inhabited by digital machines. We visit this digital world by peering through the portholes of our computer screens. . . . Now we are handing sensory organs and manipulators to the machines and inviting them to enter analog reality. The scale of possible surprise that this may generate over the next several decades as sensors, lasers, and microprocessors co-evolve is breathtakingly uncertain.

MANY SEE SPAM AS PRIVACY INFRINGEMENT AS WELL

Sticking with the theme of commercial interests a prime cause of consumers' loss of privacy in the digital age, another problem of great scale is "spam"—also known as unwanted e-mail from unwelcome purveyors. The term was inspired by a "Monty Python" sketch set in a café which only serves Spam luncheon meat—when a customer requests something else, the server repeats the Spam-packed menu and is soon joined by a chorus of Vikings chanting and singing "Spam, Spam, Spam," and drowning out all other communication.

Unsolicited commercial e-mail is seen as intrusive and an interruption of the flow of desired communication. In the 1980s, when early Internet users frequented bulletin boards and multi-user domains, some abusive users would write the word *SPAM* a number of times to force other respondents' text off the screen—this was also called "trashing" or "flooding." Later, on Usenet, *spam* was used to refer to excessive multiple postings of the same message.

The first commercial spam was sent by the law firm Canter and Seigel in 1994; the lawyers employed bulk Usenet postings to advertise the firm's immigration law services. The firm went on to promote the use of such postings by other companies of all types as a commercial-sales tactic. By 1995, commercial e-mails became the predominant form of spam. Some experts today estimate that spam may constitute as much as 80 percent of Internet mail traffic. E-mail isn't the only target of spammers; they also send unwanted mail—known as spIM—to people's IM accounts.

Microsoft founder Bill Gates understands the spam challenge; his co-worker Steve Ballmer told the audience at a 2004 technology forum that Gates gets as many as 4 million e-mails a day—most of them spam.

Gates wrote in a 2003 Microsoft "Executive E-mail" column on the company's website:

Spam is a drain on productivity, an increasingly costly waste of time and resources for Internet service providers and for businesses large and small. It clogs corporate networks, and is sometimes a vehicle for viruses that can cause serious damage. Spammers often prey on less sophisticated e-mail users, including children, which can threaten their privacy and personal security. And as everyone struggles to sift spam out of their inboxes, valid messages are sometimes overlooked or deleted, which makes email less reliable as a channel for communication and legitimate e-commerce. Spam is so significant a problem that it threatens to undo much of the good that e-mail has achieved. . . . Efforts across many fronts should lead to a world where we are less troubled by spam.

MORE PRIVACY, IDENTITY, AND ANONYMITY
STATEMENTS FROM 1990 TO 1995

Anonymity online no longer seems possible, as it did in the early days of the Internet. The statements of concern in the early 1990s have not been eased in the ensuing years. Following are some of the most compelling statements made by stakeholders at a time when privacy began to become a thing of the past.

Computer monitoring challenges traditional expectations of privacy, exposes nearly every facet of an individual's life to potential public view and commercial use, alters the relationship between employers and employees, and opens the way for unprecedented government surveillance of citizens. . . . As technology makes its easier to match databases and repackage personal information in commercially valuable forms, unease increases over the amount of information gathered and retained, where it comes from, how accurate it is, what use is made of it, and how individuals can control that use, especially when it is reused. Again, computers exacerbate the problem because they create a pervasive and long-lasting information trail that is decreasingly under the control of the individual involved.—**Nan Levinson,** in a 1992 posting on the Electronic Frontier Foundation site titled "Electrifying Speech"

Because the Net allows extremely flexible means of identifying message senders, it will facilitate the creation of pseudonymous and anonymous messaging that brings into question long established notions of human identity. We will be free (within limits) to put on and take off various personae. We will, of course, need to rethink basic questions about authenticity, identity and responsibility, in this context.—**David R. Johnson,** a Washington, D.C., lawyer and chairman of the EFF who helped draft the Electronic Communications Privacy Act, wrote this as the ninth of 10 hypotheses on the Internet and law in 1992

In time, high-tech snooping and databanking could make earlier-generation activities seem naively old-fashioned, as innocent as child's play. When that occurs, our failure to legislate controls over surveillance equipment as they evolved—already a problem today—could overwhelm us, as could our failure to prescribe adequate civil and criminal penalties for abuses of individual privacy committed by government agencies and U.S. corporations.—**Jeffrey Rothfeder,** in his 1992 book *Privacy for Sale*

As computer and network technology changes, lots of words and concepts will change their meanings. This is both inevitable and reasonable. But it also means that we will have conflicts about what words ought to mean. "Privacy" is one of those words, and we should be vigilant in defining and defending an expansive understanding of it.—**Phil Agre** in the June 1994 issue of *Network Observer,* his online newsletter

The rate of technological change will render privacy obsolete. During the critical period in which we can prevent the destruction of privacy, we cannot proceed on the assumption that those with power share our views and can be counted on to preserve our values. Brandeis saw the pitfall in such hopes as well when he said that "the greatest dangers to liberty lurk in insidious encroachment by men of zeal, well-meaning but without understanding." We must understand, and we must act.—**Frank Tuerkheimer,** in a 1993 *Communications of the ACM* article titled "The Underpinnings of Privacy Protection"

I look forward to a time when no one will be exempt from surveillance. So long as corporations, governments, and citizens are equally vulnerable, lack of privacy will be the ultimate equalizer. It will also drastically reduce crime—especially street crime—when there's a constant possibility of electronic evidence turning up in court. On a domestic level, I doubt that this will affect us much, one way or the other. If my neighbors don't bother to bug my phone right now, why should they bother to video my apartment in the future? On a macro level, the impact will be significant; and I believe most of it will be positive.—**Charles Platt,** in a 1993 article for *Wired* magazine headlined "Nowhere to Hide"

A pretty good society needs more than just anonymity. An online civilization requires online anonymity, online identification, online authentication, online reputations, online trust holders, online signatures, online privacy and

online access. All are essential ingredients of any open society.—**Kevin Kelly,**
Wired magazine editor, in his 1994 book *Out of Control: The New Biology of
Machines, Social Systems and the Economic World*

Everything we read, spend money on, or do will literally be a database. . . .
Many people believe that privacy will be the nightmare issue of the NII.
—**Fred Weingarten,** executive director of the Computer Research Association,
in a 1994 *Business Week* article titled "From Internet to Infobahn"

It has been thoughtlessly said . . . that cryptography brings the unprecedented
promise of absolute privacy. In fact, it only goes a short way to make up for
the loss of an assurance of privacy that can never be regained.—**Whitfield
Diffie,** in a 1994 *Wired* article titled "Prophet of Privacy"

Computerized data banks empower bureaucratic authorities by providing
easy access to personal information—about credit ratings, school performance, housing, medical histories, and tax status. And in the future, they will no
doubt allow access to genetic profiles, providing information about our pre-
disposition to certain diseases or behavioral conditions. Such information
may be available to employers, insurers, product advertisers, banks, school
systems, university tenure committees, and other institutions that exercise
enormous control over our lives. Indeed, given its social impact, computerization could well be called the "cursor" of our time.—**Dorothy Nelkin,** in a
Spring 1994 *National Forum* article titled "Information Technologies Could
Threaten Privacy, Freedom, and Democracy"

Anonymity is scary stuff. Quite reasonably, administrators worry about terrorists claiming credit for bombings or kidnappers posting ransom notes.
Even in their experimental infancy, remailers have had an effect on Net culture. On one hand, there's been an outburst of harassing or simply idiotic
flames that have no return headers. On the other hand, there's now a way for
victims of sexual crimes or whistle-blowers to send mail and messages with
the assurance of privacy. At this point, however, probably the most important
role that remailers play is to launch a necessary dialogue on the issue of
anonymity in a digital society. What are the benefits and the risks? Even if we
don't want it, can we stop it?—Technology writer **Steven Levy,** in a 1994 *Wired*
magazine article titled "Anonymously Yours"

What about selling information that isn't viewed as legal, say about pot grow-ing, do-it-yourself abortion, cryonics, or even peddling alternative medical in-formation without a license? What about the anonymity wanted for whistleblowers, confessionals and dating personals? . . . Encryption always wins.—Cryptographer **Tim May,** quoted in Kevin Kelly's 1994 book *Out of Control*

Privacy technology, whether used for electronic payments, voting, or other public expression, is the electronic equivalent of a free market and democracy. People will come to insist on it as an informational human right.—**David Chaum,** founder of DigiCash and cryptography expert, in his testimony at a 1995 congressional hearing

Certainly the Internet didn't invent threats, defamation, copyright infringe-ment and other illicit speech. But with its anonymous remailers, the Net does enable speakers to avoid apprehension and punishment for such illegal utter-ances. Banning online anonymity would go much too far, stifling the 99 per-cent of legitimate speech in pursuit of the 1 percent of criminal speech.—**Stephen Bates,** a senior fellow at the Annenberg Washington Pro-gram, in a 1995 article for *CQ Researcher* titled "Regulating the Internet"

You deserve at least as much anonymity on the Net as you have when you cast a vote, post an anonymous tract, or buy a newspaper from a coin-operated rack. In fact, you should demand a stronger right on the Net. Otherwise, au-thorities will find it easy to track, sort, and record your digital behavior. You should thus demand the right to use the most powerful encryption available. Uploading a robust right to anonymity calls for public key cryptography.—**Tom Bell** of the University of Dayton Law School, in a 1995 *Wired* magazine article titled "Anonymous Speech"

Electronic coding of messages that will permit freedom of choice to deal with anonymous messages or that will refuse to deal with them should be devised. Guidelines should be refined so as to permit the use of aliases and pseudo-nyms in electronic playgrounds and to preserve the privilege of posting anonymous messages when doing so serves some useful public purpose like whistle-blowing. To preserve order and civility, however, abusive posters of anonymous messages must not be permitted to insulate themselves from accountability for their wrongdoing.—**Anne Wells Branscomb,** an expert in technology and the law, in a 1995 *Yale Law Journal* article titled "Anonymity, Autonomy, and Accountability"

Even if you don't cruise the superhighway, your personal profile will. A portrait of you in 1's and 0's, the language of computers, will exist in cyberspace. The profile could be so complete that it will be like having another self living in a parallel dimension; it is a self you cannot see, but one that affects your life just the same. Even if you do not own a personal computer and never intend to, you are part of the revolution. . . . The privacy problems posed are so different than those that have come before, there is no framework to deal with them. Technology is fast. The law, whether formed in tiny increments by individual cases or by the cumbersome legislative process, is slow. As a result, there is simply no comprehensive body of law established to deal with all of the privacy concerns arising in the digital age.—**Ellen Alderman and Caroline Kennedy,** in their 1995 book *The Right to Privacy*

Hmmm . . . Will It Happen?

These predictions did not come true; nor do they seem likely to come to pass; then again, you never know

When you are trying to figure out the future, it's understandable you will sometimes project things that just don't come to pass in the way you had expected. As Steven P. Schnaars pointed out in his 1989 book *Megamistakes: Forecasting and the Myth of Rapid Technological Change,* a number of factors conspire to cause errant projections. According to Schnaars, these include: the tendency for forecasters to be personally smitten with the technology, ignoring the market it will serve; a bias toward optimism about new technology (despite the fact that most new products fail, there is always an enthusiastically convincing support structure behind their production); and the mistaken assumption that the issues and political and social concerns of the past will remain the issues and concerns of the future.

For instance, in the late 1960s most people, convinced by the success of the missions to explore the moon, were certain that Earthlings would send a manned mission to Mars before the year 2000. "A Report on Tomorrow," published by National Underwriter in the late 1960s, claimed that the 1980s would see underwater hotels, orbiting space factories, and pre-built houses delivered by helicopter. It could be that predictions about a robot takeover, or the development of biotechnology to the point in which a "godmind" runs the planet, will eventually wind up on the discard pile of failed prophecies. We certainly don't know, but for now these are intriguing prospects.

IN RETROSPECT, POOR PROGNOSTICATIONS
CAN ELICIT GIGGLES OR GUFFAWS

Failed predictions can sometimes seem quite humorous in retrospect, but their predictors are sincere, and their words reflect the thinking of their times. While some people of the early 1990s went out on a limb regarding the specifics of the future of the Internet, most hedged their predictive statements well enough that their statements weren't laughable a decade later. Perhaps those voices predicting the future of networked communications learned a lesson from the following folks of the past and recent past whose forecasts became their follies. Classically inaccurate guesses at the future include the following:

A man has been arrested in New York for attempting to extort funds from ignorant and superstitious people by exhibiting a device which he says will convey the human voice any distance over metallic wires so that it will be heard by the listener at the other end. He calls the instrument a telephone. Well-informed people know that it is impossible to transmit the human voice over wires.—News item from a New York paper, circa 1868

This "telephone" has too many shortcomings to be seriously considered as a means of communication. The device is inherently of no value to us.—Western Union internal memo, 1876

Louis Pasteur's theory of germs is ridiculous fiction.—**Pierre Pachet,** professor of physiology at Toulouse, France, 1872.

Heavier-than-air flying machines are impossible.—**Lord Kelvin,** president, British Royal Society, 1895

Everything that can be invented has been invented.—**Charles H. Duell,** commissioner, U.S. Office of Patents, 1899

God himself could not sink this ship.—Deckhand on the *Titanic,* April 10, 1912

The wireless music box has no imaginable commercial value. Who would pay for a message sent to nobody in particular?—RCA executive **David Sarnoff's** associates in response to his urgings for investment in radio in the 1920s

Who the hell wants to hear actors talk?—**H. M. Warner** of Warner Brothers film studio, 1927

Stocks have reached what looks like a permanently high plateau.—**Irving Fisher,** professor of economics, Yale University, just before the crash in 1929

There is not the slightest indication that nuclear energy will ever be obtainable. It would mean that the atom would have to be shattered at will. —**Dr. Albert Einstein,** 1932

The energy produced by the atom is a very poor kind of thing. Anyone who expects a source of power from the transformation of these atoms is talking moonshine.—Physicist **Lord Ernest Rutherford,** 1908 Nobel winner, after splitting the atom for the first time, 1933

I think there's a world market for about five computers.—**Thomas Watson,** chairman of IBM, 1943

The bomb will never go off, I speak as an expert on explosives.—**Admiral William Leahy,** U.S. atomic bomb project, 1945

Where a calculator on the ENIAC is equipped with 18,000 vacuum tubes and weighs 30 tons, computers in the future may have only 1,000 vacuum tubes and perhaps weigh 1.5 tons.—Unlisted author, *Popular Mechanics,* March 1949

It would appear that we have reached the limits of what it is possible to achieve with computer technology, although one should be careful with such statements, as they tend to sound pretty silly in five years.—**John Von Neumann,** 1949

I have traveled the length and breadth of this country and talked with the best people, and I can assure you that data processing is a fad that won't last out the year.—The editor in charge of business books for Prentice Hall, 1957

We will bury you!—**Nikita Khrushchev,** Soviet Union premier from 1958 to 1964

We don't like their sound, and guitar music is on the way out.—Decca Recording Co. executive rejecting the Beatles, 1962

But what . . . is it good for?—Engineer at IBM, 1968, commenting on the microchip

Get your feet off my desk, get out of here, you stink, and we're not going to buy your product.—**Steve Jobs's** account of how Joe Keenan, president of Atari, in 1976 responded to Jobs's offer to sell him rights to the new personal computer he and Apple Computer co-founder Steve Wozniak developed

There is no reason for any individual to have a computer in his home.—**Ken Olsen,** President of Digital Equipment Corporation, at the Convention of the World Future Society, 1977

The 32-bit machine would be "overkill" for a personal computer.—**Sol Libes,** electrical engineer, in the column ByteLines in *Byte* magazine 1981

640K ought to be enough for anybody.—**Bill Gates,** 1981

We don't see Windows as a long-term graphical interface for the masses.—Lotus Development official, demonstrating a new DOS version of Lotus 1-2-3, 1989

NONE OF THESE GO TO THOSE EXTREMES, BUT THEY ARE WRONG

Following is a selection of predictions about the Internet, culled from thousands of statements made between 1990 and 1995, that were made by people trying to get a handle on the future. These projections do not seem likely to come to pass. Some of them are absolutely wrong. Those in which there is no time span listed still have a chance to come to fruition. Many of these statements were made as generalized conjectures rather than determined diatribes; their authors were merely surmising about possibilities or taking an exaggerated stance to make a point that still has or had some merit.

Prediction: People could just form co-ops and fiber themselves up and have a neighborhood node supplied by the lowest bidder, instead of waiting for some centralized force to come in and lay fiber. —Virtual-reality entrepreneur **Jaron Lanier**, quoted in a 1995 article for *Engineering Times* titled "The Nightmare Scenario"

Explanation: This proposal was a venting of frustration over the seemingly endless haggling over what the Internet delivery system should be. The idea was that people didn't need to wait for industry to build all of the neighborhood tie-ins to the network infrastructure; instead, people would do it. In the years following this statement, cable companies and phone companies improved their ability to deliver Internet connections of varying speeds. As wireless technologies began to blossom, some neighbors did begin to share their own access to the Internet, intentionally or otherwise.

Prediction: To my mind, it is likely that what we now understand as the mass media will be gone within 10 years. Vanished, without a trace. . . . Who will push *The New York Times?* The answer, I think, is technology. . . . Consumers will naturally want better information. They'll demand it, and they'll be willing to pay for it. —Author **Michael Crichton**, in a 1993 article for *Wired* titled "Mediasaurus: Today's Mass Media Is Tomorrow's Fossil Fuel"

Explanation: Crichton's criticism of the media came in an essay for *Wired* magazine. It triggered a useful discussion, but the headline-grabbing idea that traditional mass media formats would disappear in the next decade was completely off. In fact, many old-school media outlets began online versions that are quite successful, while they also retained their original format in basically the same old form.

Prediction: The telcos might even create an alternate Internet, one that bills by traffic instead of bandwidth. They would then market this alternative as the "uncongested" Internet where applications like Internet telephony could reach their full potential.—**Danny Briere**, telecommunications executive, in a 1995 *Wired* article headlined "IPhone: Will Telephony on the Internet Bring the Telcos to Their Knees?"

Explanation: Thus far, bandwidth has grown to keep pace with consumer demand, and the only "alternate Internet" is Internet2, for researchers' needs. The top forms of Internet communication are e-mail and instant messaging. Cellular phones are far more popular than Internet telephony.

Prediction: [There is a] need for a simple e-mail address system that gives every U.S. resident a "default" e-mail address by which they can be reached.

Such a development would "jump start" a universal access system, because governmental and other organizations could then assume that "everyone" was reachable by this means and design procedures and systems accordingly. . . . A simple e-mail address-provision scheme should be developed giving every U.S. resident an e-mail address, perhaps based on a person's physical address or telephone number.—Researchers **Robert Anderson, Tora Bikson, and Sally Ann Law,** in their 1995 paper "Universal Access to E-mail," which is part of the RAND Publications Database
Explanation: The government has not assigned every U.S. resident a default e-mail address. Individuals prefer independence and the possibility of anonymity, so it is unlikely that such a proposal would meet with success if put into practical use. Some institutions (universities, corporations) assign each member an e-mail address; often the members do not read the mail that comes to that address, leaving it to pile up to the point at which the mailbox is full and mail cannot be delivered.

Prediction: I don't . . . think that my telephone will merge with my computer, to become some sort of information appliance.—Physicist and Internet critic **Clifford Stoll,** in the introductory chapter, titled "A Speleological Introduction to the Author's Ambivalence," in his 1995 book *Silicon Snake Oil*
Explanation: The telephone has not only merged with the computer, it is merging with cameras, MP3 players, video game software, GPS systems, you name it. And, yes, it is definitely an information appliance.

Prediction: The fear that individuals will be overwhelmed by a deluge of "junk mail," or subjected to defamatory or otherwise inappropriate message content, appears to be, although not a trivial issue, at least not one requiring too much attention at this point.—Researchers **Robert Anderson, Tora Bikson, and Sally Ann Law,** in their 1995 paper "Universal Access to E-mail," which is part of the RAND Publications Database
Explanation: Spam, anyone? Unwanted electronic mail, some of it with inappropriate content, is a major public issue of some concern. It has drawn a great deal of interest and attention.

Prediction: The Internet as we know it will be passe in five years—just as the largest number of people are waking up to it and making investment decisions about it. They will soon look foolish.—Montana computer bulletin board host **Dave Hughes,** in a 1994 *Wired* article titled "Chaos Is the Form"

Explanation: The Internet has gained quite a bit of ground since the mid-1990s. It has never been passé. However, those who made the wrong investment decisions did wind up looking foolish if they didn't sell at the right time, so this could be seen as partly correct.

Prediction: By the year 2000, long-distance video conversations will be a commonplace feature of the Information Superhighway.—Internet pioneer **Lawrence Landweber,** quoted in a 1994 *Wisconsin State Journal* article titled "The Internet: Computer Network Is Superhighway On-ramp"
Explanation: The video aspect was present, it just didn't really arrive as a popular form of Internet communication by 2000. Some people did enjoy long-distance video conversations via the Internet, but they were not a commonplace feature yet.

Prediction: The Wintel (Windows/Intel) personal computer is "ridiculous"—too complicated to use and too expensive for the average user.—Paraphrase of remarks Oracle CEO **Larry Ellison** made in 1995, as reported in a *Context* magazine article in 1999 titled "The Five Worst Predictions of the Internet Age"
Explanation: The Wintel system is the dominant computer in use in businesses and homes. Because it cost less than Apple computers and because the industry chose to stick with the Windows/Intel format, it has remained vibrant into the 21st century.

Prediction: In the future, every Internet operator will be subject to local laws. . . . And software will be developed to provide the appropriate local censorship.—**Eric Schmidt,** chief technology officer for Sun Microsystems, quoted in a 1995 *New York Times* article titled "The Media Business: Online Service Blocks Access to Topics Called Pornographic"
Explanation: At this point, no locality-specific censorship software has been developed for use on the Internet in the United States. Some other nations with closed political systems endeavor to exercise control over Internet uses.

Prediction: There will be a trade association that will publish a ratings system that Prodigy, Apple, Microsoft, etc. all agree on. They will subscribe to it, all netwatch software comes with it by default, and, here's the kicker, anyone who doesn't support it runs the risk of being hauled into court.—**Seth Finkelstein,**

computer programmer and anti-censorship activist, in a 1995 Electronic Frontier Foundation essay

Explanation: No particular Internet ratings system has been universally adopted, although companies have evolved software to try to help parents and schools filter inappropriate material out. It has been difficult to develop software that does not also block out useful materials, such as information on AIDS prevention or support sites for breast cancer victims.

Prediction: Digital credentials . . . reveal no unnecessary information, people would be willing to use them even in contexts where they would not willingly show identification. . . . They may also acquire negative credentials, which they would prefer to conceal: felony convictions, license suspensions or statements of pending bankruptcy. In many cases, individuals will give organizations the right to inflict negative credentials on them in return for some service. —David Chaum, DigiCash founder, in a 1992 article for *Scientific American* titled "Achieving Electronic Privacy"

Explanation: It may be too soon to tell what will happen in regard to this prediction. Chaum says people will be willing to carry digital credentials that show negative information—for instance, that they are convicted criminals. This has not happened, and it seems unlikely to win public support in a democracy such as the United States.

Prediction: In a digital world, the bits are endlessly copyable, infinitely malleable, and they never go out of print. . . . Pass a Bill of Writes—a digital deposit act—requiring that each item submitted to the Library of Congress be accompanied by its digital source. Make it illegal to obtain copyright otherwise. . . . Instead of being the "library of last resort," it might become the first place to look. . . . A Library of Progress could be in the pockets of tomorrow's kids. Having a Bill of Writes now means that we can spend the next 20 to 50 years hammering out new digital-property laws and international agreements without stunting our future.—Nicholas Negroponte, in a 1995 *Wired* column titled "A Bill of Writes"

Explanation: It is unlikely that people will be required to submit every copyrighted item to the Library of Congress. This solution was never taken seriously—was it meant to be? Copyright remains a largely unresolved issue in networked communications.

Prediction: With the advent of digital cash, anyone can set up their own stories-serving station—hundreds upon thousands of little self-sufficient magazines, supported by communities of folks who care to share. . . . Reporters will then leave major media, set up their own content-serving stations. . . . We'll all forget about money, sit around and make art, and tell each other stories, while computers handle all the problems of the world. Why not? —Internet enthusiast **Justin Hall**, in a 1995 speech for a conference titled "News Industries & Journalism: Preparing for 2010"

Explanation: Despite the hype for blogs, most reporters are still enjoying their fringe benefits (paid vacation, health insurance) and working for the major media outlets. Few people have found the arrival of the Internet has freed them up to "sit around" while computers handle all the problems of the world. In fact, many people would say that computers have increased their workload exponentially.

Prediction: As payments on the network mature, you're going to be paying for all kinds of small things, more payments than one makes today, and they're going to be that much more revealing. Every article you read, every question you have, you're going to have to pay for it.—**David Chaum**, 1994, in a *New York Times* article headlined "Attention Internet Shoppers: E-cash Is Here"

Explanation: The idea of micropayments for every piece of information gained was popular in the mid-1990s. But entrepreneurs underestimated the stubborn desire of the Internet culture; people refused to pay for content, avoiding fee-based sites in general and continuing to flock to free information. Few fee-based systems have been successful.

Prediction: If you're not an active Internet citizen by the mid-1990s, you're likely to be out of business by the year 2000.—Computer consultant **Patricia Seybold**, 1994, in a *Computerworld* article that was quoted in a *New York Times* article headlined "Getting Down to Business on the Net"

Explanation: Many businesses have become more profitable thanks to the Internet. But plenty of businesses have survived quite well with no Internet presence. Seybold was just exaggerating to make a point; but commerce has not taken hold to the extremes that people in the early 1990s were predicting it would—yet.

Prediction: Internet telecommuting will become commonplace within larger businesses as groupware enables management and professional employees to achieve "virtual proximity" while working at home part-time.—Technology writer **Paul Crisci,** in a 1995 *Red Herring* article titled "The Internet Is Much More Than an Information Highway"

Explanation: While it is a definite part of our work culture, telecommuting by U.S. workers to their traditional jobs has not become as popular as some expected it to be. Why? People found the need to be involved in their businesses' culture. They did not like being out of personal touch with co-workers—feeling out of the loop when working at home. Some found it difficult to work from home due to the demands they felt in that atmosphere. By the year 2000, telecommuting numbers were actually dropping off a bit from the figures earlier in the 1990s. Teleworking numbers, however, were up, as American companies outsourced some jobs to workers in other nations to save on costs. For instance, many U.S. software companies are hiring programmers based in India. From 2000 to 2001 the total number of teleworkers jumped 17 percent. Another burgeoning development in the field of telework is the extended workday: many workers are now putting in a full day or more at the office and then taking their work home with them, thus extending their typical workdays to a great degree—this is by far the most regularly occurring telework taking place in the United States.

Prediction: With the technological face-lift they're giving the whole White House system, it won't be long before you can comment on a policy and get a personal reply.—Technology developer **Brendan Kehoe,** in a 1993 talk at a bookstore to promote his book *Zen and the Art of the Internet*

Explanation: Write an e-mail and get the immediate attention of the president's advisers—it's digital democracy at work. Maybe not. It has been found that e-mail gets to the White House fast, but it likely gets no more attention than old-fashioned snail mail. In fact, because of the cut-and-paste nature of some e-mail campaigns and the anonymity afforded by the Internet, e-mail can sometimes even have less weight than a traditional letter hand-written and hand-signed by you and delivered by the U.S. Postal Service.

Prediction: Although the use of electronic mail may sound more promising, fax has a better foothold now, and it may be 10 years before something better replaces it.—Software innovator **Dan Bricklin,** quoted in a 1990 *InfoWorld* article titled "PC Fax Modems, OCR Software, Scanners Growing in Popularity"

Explanation: Bricklin really liked his fax machine. Do you think he still does? E-mail blew faxes away pretty fast in the 1990s, and instant messaging has also done its part to replace the fax. Why waste all that paper? People can easily scan signed documents or pictures and send them as e-mail attachments on the Internet.

Prediction: The excitement that you get at an auction and the fever that you get that drives the prices up, for instance, I just don't think you can duplicate on-line.—**Cathy Sykes,** vice president of the Show Managers Association, quoted in a 1995 *Norfolk Virginian-Pilot* article headlined "The Future of Antiques"
Explanation: eBay. Enough said.

Prediction: The era of public-access Internet has come to an end.—*Wired* magazine publisher **Louis Rossetto,** in a 1995 statement to *Wired* employees quoted in Gary Wolf's 2003 book *Wired: A Romance*
Explanation: Rossetto was a businessman who thought he could make big profits selling content online. He misunderstood the continuing influence of the culture built into networked communications by its originators—people who shared information freely. The public-access Internet remains, with a great majority of the information offered at no directly billable cost to users. Few people pay for content.

Prediction: Network providers can use pricing and other means to balance load and to shift usage to off-peak periods. In addition to congestion-related variation in pricing based on time of use . . . another timing issue relates to making long-term versus spontaneous arrangements to use above-average amounts of network capacity. Users who anticipate only occasional demand for very-high-speed service can be encouraged to make an explicit reservation in advance.—NRE-Naissance Committee, appointed by the Computer Science and Telecommunications Board of the National Research Council, in its 1994 report
Explanation: There is no need to make reservations to use the Internet—isn't that nice? This prediction was made at a time when bandwidth was at a premium and people who used the Internet had to wait their turn to send and receive information. Those bad old days disappeared thanks to the inventive engineers and computer scientists who found ways to make enough bandwidth for everyone.

Prediction: How can use be rationed? Economists Jeffrey MacKie-Mason and Hal Varian at the University of Michigan propose a "congestion pricing" scheme. When your computer sent off a packet, it'd attach a note of how much you were prepared to pay for it to arrive promptly—say 0.002 U.S. cents for 200 bytes. Automatic real-time "auctions" at switching centers would determine the lowest "bid" at which packets would be passed; every packet which got through would pay that lowest price, and others would be delayed. When there was no congestion, all packets would travel free. . . . "Netizens" really would become hitchhikers on the infobahn, squeezing into the gaps in the coming flood of commerce.—Technology writer **Mike Holderness** explains a proposal from University of Michigan economists **Jeffrey MacKie-Mason and Hal Varian,** in a 1994 essay in *The (London) Guardian* titled "High Anxiety for Hitchhikers on the Infobahn"

Explanation: This proposal also originates in the days of limited bandwidth. Thanks to new developments in the technology, people were never forced to decide how much they would be willing to pay to send their e-mail.

Prediction: The change in the technology is so dismaying to so many people, to so much business that there may be an attempt to just lock it in and say, "OK, Bill Gates, run the world. Everything's going to be Microsoft, everything's going to be on the Net, and the computers—that's it. Let's have one standard and slow things down so versions come at a predictable pace." There's not a lot of competition, which drives things too fast, it's too confusing. Let's just make it stable. That would be people asking for a monopoly in order to manage creative technological change.—Internet pioneer **Stewart Brand,** in a 1995 interview that aired as part of the PBS-TV program *High Stakes in Cyberspace*

Explanation: Brand was speculating and exaggerating to make the point that things were moving so fast it was scary. Most people—especially Brand— would likely say they are pleased that they didn't just agree to let Bill Gates or any other executive or single corporation "run the world."

NICHOLAS NEGROPONTE VERSUS GORDON BELL: ONE FINAL ESTIMATE GONE AWRY

When I was working on the research for this book late in 2004, I had an interesting e-mail exchange with Microsoft's Gordon Bell. He gets a kick out of being a guy who has been at the cutting edge during the most fascinating decades in the history of network technology, and he wanted to share the details of a 1990s bet he made with Nicholas Negroponte of MIT.

"John Perry Barlow would not take this bet," he wrote. "It was just with Negroponte. It was made in 1997 and to be decided Jan. 1, 2001, and 2002."

Negroponte and Barlow both expected the growth of new users on the Internet to continue at a fantastic rate through the late 1990s and into the new century. In a 1998 e-mail he sent to a third party along with a copy to Bell, Negroponte shares the fact that he made a bet with Bell in which he estimated that "there will be 1 billion" users on the internet "before the end of the year 2000" (measured by counting individuals, not by e-mail addresses, since people sometimes have multiple addresses).

Bell has kept the e-mail evidence to prove that Negroponte took the bet. He says, "Negroponte has still not paid off, and I have cited dozens of surveys since then that give at most 750 million users as of 2003. I would like to see this put on the record!"

Appendix A: *Wired* Inspired

The incredible influence of a pulp-based product

From its first days in print in 1993, *Wired* magazine initiated controversy. Many considered its initial editorial tone to be arrogant and self-important, but it was also seen by some as the high-tech-age equivalent of *Rolling Stone,* a publication that carried significant influence as the noisy voice of the rock 'n' roll revolution of the 1960s.

Love it or not, *Wired* packed a wallop, carrying stories that influenced key decision makers in the explosive early years of the public Internet—the 1990s. There is no doubt that *Wired* was the stage on which some of the most important topics of the time were brought forth for spirited debate. Because the magazine looked forward, its content provided a great deal of the predictive material published in this book and in the Elon University/Pew Internet Predictions Database, a record of internet predictions made between 1990 and 1995. *Wired* thus deserves a closer look, and this appendix is a tribute to its writers and editors and their contributions as illuminators in the new network age. The complete archives of *Wired* are available online for free, and they are definitely worth your attention and further study.

Founding editor and publisher Louis Rossetto, in explaining his ideal for *Wired* to newly hired executive editor Kevin Kelly in 1992, said the magazine should be like a knowing missive sent back from the future. In the book *Wired: A Romance,* author and former *Wired* staffer Gary Wolf quotes Kelly: "*Wired* would be like a reverse time capsule. It would sail back through time and land at people's feet. . . . They would be perplexed, fascinated, revolted, mesmerized." Wolf also writes that Kelly used a quote from cyberpunk author William Gibson as a motto for *Wired*: "The future is here—it's just unevenly distributed."

ROSSETTO AND METCALFE MADE SACRIFICES TO GET IT GOING

Previously teamed in the 1980s in the operation of an Amsterdam-based publication, *Electric Word, Wired* co-founders Louis Rossetto and Jane Metcalfe spent several years scrambling to get their new U.S.-based publication off the ground.

Rossetto had described the readers of *Electric Word* as "thinking computer users involved in the global revolution of information technology." The publication never had more than 15,000 subscribers, it had disappeared by 1990, but their work on *Electric Word* reinforced the team's belief that there was a need for a new magazine that could document the coming technology issues and influence change. Metcalfe came up with the name *Wired,* plans were made, and financing was found with a great deal of difficulty over the span of the next few years.

Wolf and other tech community members of the time tell about a 16-page sample publication that was cobbled together in 1991 to attract financial backing for *Wired.* In it, Rossetto wrote, "You, the information-rich, are the most powerful people on the planet today. You and the information technology you wield are completely transforming our lives, our families, our neighborhoods, our educations, our jobs, our governments, our world."

Rossetto and Metcalfe took what they called issue number zero of *Wired* to the third TED (Technology, Entertainment, Design) Conference in February 1992, where they showed it to Nicholas Negroponte, the influential communications-technology booster who had recently helped gain the financing to build the Massachusetts Institute of Technology's Media Lab. He pledged his personal support of the publication, plus $75,000, and a monthly column for *Wired* in exchange for 10 percent ownership of the magazine. His approval and participation and additional funding from other people helped Rossetto and Metcalfe gain the backing they needed to keep the dream of publishing their magazine alive.

IT WASN'T A SLAM-DUNK AT THE START

There is no way to underestimate Rossetto's zeal for what he saw as his calling. *Wired* staffer Wolf describes a scene from 1992 in his book: "Louis. . . . tried to make [the first staff members he hired] understand *Wired*'s larger significance. He said the curve of history was at an inflection point, which meant that their actions today could affect human destiny far into the future. It was impossible to overestimate the importance of such moments. . . . *Wired* would advocate what Louis called 'spontaneous order,' . . . a more benign mode of human civilization that was inherent in digital technology."

Wired No. 1 went to the printer in December of 1992 and was released in January of 1993. In it, Rossetto wrote, under the headline *Why Wired?* "Because the Digital Revolution is whipping through our lives like a Bengali typhoon—while the mainstream media is still groping for the snooze button." Response to the first issue and every issue thereafter was mixed. Many tech insiders liked what they saw, but its offbeat design and bizarre use of typography made *Wired* difficult to read, drawing many complaints.

Immediately after the first issue appeared, magazine trade journal *Folio* carried an article in which it quoted analysts' opinions about the new publication. Dan Orlow of Periodical Studies Service in New York said, "I don't see it as the wave of the future. It's more of a trial balloon. Frankly, I don't think they have a prayer." Peter Craig of Magazine Consulting Group in Los Angeles said, "My impression is that people want service and information—they are not interested in the lifestyles of computer nerds." Gary Chapman, director of Computer Professionals for Social Responsibility, wrote a letter to the editor of *Wired*, complaining that the magazine was "yuppie bullshit." The letter was printed in a *Wired* Letters to the Editor column. Newsweek carried an item about *Wired* headlined: "Propeller Head Heaven: A Techie Rolling Stone."

Insider Wolf reports in his book that Rossetto and Metcalfe bounced checks and avoided creditors in order to stay in business through 1993, but the magazine was popular enough among the masses of the computer-literate to bring in some advertising and subscription revenue, and by the end of the year Conde Nast had purchased a 15 percent stake in *Wired* Holdings for $3.5 million and *Wired* had won a National Magazine Award for Excellence from the American Society of Magazine Editors.

POLICYMAKERS BECAME FANS;
THE BEST WRITERS MADE CONTRIBUTIONS

Soon White House aides and politicians' assistants were quoting information from *Wired* to demonstrate that they and their bosses were up to speed on technology and the new economy; *Wired* began appearing in waiting rooms across the nation, figuring into a sort of reflection of the cutting-edge connectedness of professionals and businessmen.

In its early years, the content of the magazine was determined by Rossetto, Kelly, and managing editor John Battelle (who later founded a rival publication, *The Industry Standard*). *Wired* lured the best writers to report and write on the most pressing issues of the time—issues that could have a profound effect on the future. Marc Andreessen, John Perry Barlow, Stewart Baker, John Browning, Esther Dyson, George Gilder, Mitch Kapor, Jay Kinney, Steven Levy, Nicholas Negroponte, Joshua Quittner, Paul Saffo, Lewis Perelman, Lance Rose, Michael Schrage, Evan Schwartz, Denise Caruso, Bruce Sterling, Bill Joy, Lawrence Lessig, Katie Hafner, Neal Stephenson, Alvin Toffler, and dozens of others voiced their concerns, ideas and enthusiasms as writers or interviewees.

The contrarians were not left out. *Wired* featured long interviews with self-proclaimed neo-Luddite Kirkpatrick Sale and with technology critic Sven Birkerts, giving them a platform from which to share their viewpoints.

In 1996, there was a failed attempt to take the magazine public. Rossetto and Metcalfe lost control of the company in 1998. Providence Equity sold it off, and it became a part of the purvey of Conde Nast, a New York-based publishing firm. It developed into a slick advertising vehicle for upscale products, each edition packed with colorful, glossy ads, but it continued to carry the sometimes irreverent or anti-tech/anti-commercialization/anti-establishment viewpoints of respected commentators such as Joy, Lessig, and Sterling.

CLASSIC *WIRED* ONE- AND TWO-LINERS

What did the articles in the issues of *Wired* published in the early 1990s tell us in their predictive style? They said this time was a turning point for the world, and they brought great ideas to light for the general public and policy makers, spurring discussion and motivating change. Among the stirring one-line or two-line quotes from many voices from *Wired*, all stated somewhere between 1993 and 1995, are:

Trusting the government with your privacy is like having a Peeping Tom install your window blinds.—**John Perry Barlow**

The benefits to be had are so great; we just have to be sure that the people who are in control respect our privacy.—**Roy Want**

Technology is close enough to being out of control that human intervention has become a weaker and weaker constraint.—**Vernor Vinge**

I think that our existing political and moral structures are going to explode. —**Alvin Toffler**

You could steal a song [online], but who could steal them all? And if the listening fee is low enough, no one would bother to make copies.—**Ken Thompson**

Countries that now have lousy oppressive governments and smart, determined terrorist revolutionaries are gonna have lousy oppressive governments and smart determined terrorist revolutionaries with computers.—**Bruce Sterling**

Video conferencing bears a terrifying promise: Distance will no longer be an excuse for not attending meetings.—**Steve Steinberg**

We seem to be creating, through media and communications technology, what some have called a species-wide nervous system.—**R.U. Serious** (real name, Ken Goffman)

There are two moral judgments against computers: One is that computerization enables the large forces of our civilization to operate more swiftly and efficiently in their pernicious goals of making money and producing things. . . . And secondly, in the course of using these, these forces are destroying nature with more speed and efficiency than ever before.—**Kirkpatrick Sale**

If we were stuck with having to make technology that was centralized and stupid and brute, we would be looking forward to a dismal future. But we don't have to make technology that way. . . . In the end, people will choose technology and civilization.—**Kevin Kelly**

Now the "vast wasteland" . . . will be supplanted by a vaster wasteland brimming with utterly new forms of interactive cyberdreck. . . . The future belongs to neither the conduit or content players, but those who control the filtering, searching and sense-making tools.—**Paul Saffo**

Without junk, there is less of a chance for real quality to emerge. Let the marketplace of ideas rule.—**Mitchell Kapor**

The robots will re-create us any number of times, whereas the original version of our world exists, at most, only once. Therefore, statistically speaking, it's much more likely we're living in a vast simulation than in the original version. —**Hans Moravec**

Grrr. Machines disappoint me. I just can't love any of these wares, hard or soft. I'm nostalgic for the future. We need ultrahigh res! Give us bandwidth or kill us!—**St. Jude** (real name, Judy Milhon)

It would be easier to get the Pope to become a Buddhist than to get the schools to change.—**Ed Lyell**

Perhaps it will be fitting justice if the catalysts of the new digital politics are ultimately forced by the logic of their political ideals to become online Dr. Frankensteins battling their own creation run amuck.—**Jay Kinney**

Today's media are what the Net should never become—but will surely evolve into if it fails to develop, articulate, fight fiercely for, and maintain a value system other than expanded memory, whiz-bang toys, and money. . . . The new generation faces enormous danger from government, from corporations that control the traditional media, from commercialization, and from its own chaotic growth.—**Jon Katz**

If we establish an information-age, economically decisive, exciting America, the rest of the world will imitate us. . . . The genius of a free society is that if you are in fact correct about your perception of the future, people will aggregate and reinforce it.—**Newt Gingrich**

Your snooty little view of the world and "aren't-I-cool" sentiments are about to go crashing down on your ears. That'll piss a lot of people off, but that's good! And I say, Good riddance to the old Internet.—**George Colony**

We are entering a new economic environment—as different as the moon is from the earth—where a new set of physical rules will govern what intellectual property means, how opportunities are created from it, who prospers and who loses. Chief among the new rules is that "content is free."—**Esther Dyson**

Many of the most promising visions of how to coordinate the far-flung communication and computing cycles of this emerging platform converge on a controversial solution: the use of . . . free-ranging, self-replicating programs, autonomous Net agents, digital organisms—whatever they are called, there's an old-fashioned word for them: computer viruses.—**Julian Dibbell**

It is likely that what we now understand as the mass media will be gone within ten years. Vanished, without a trace. . . . Who will push *The New York Times?* The answer, I think, is technology.—**Michael Crichton**

If somebody comes along who's a persuasive demagogue and really commands the wires, commands the codes, and is an irresistible presence, I can imagine some dangerous scenarios. . . . Electronic church or electronic Reich? It could be either.—**Sven Birkerts**

CLEVER, DETAILED HEADLINES DEFINED *WIRED* STYLE

Wired's bold headlines between 1993 and 1995, many of them carefully edited by co-founder and publisher Louis Rossetto, attracted readers and made stirring statements of their own. A sampling reveals concerns of the day:

Prodigy, AOL, and CompuServe Are All Suddenly Obsolete—and Mosaic Is Well on its Way to Becoming the World's Standard Interface

"Anarcho-Emergentist-Republicans": Is There a New Politics Emerging in the Net/Cyberspace/Digital Culture?

Jackboots on the Infobahn: Clipper Is a Last-Ditch Attempt by the United States, the Last Great Power from the Old Industrial Era, to Establish Imperial Control Over Cyberspace

You're Not Paranoid: They Really Are Watching You; Surveillance in the Workplace Is Getting Digitized—and Getting Worse

"This Is a Naked Lady": Behind Every New Technology Is . . . Sex?

Music on Demand: Bell Labs's Ken Thompson, the Father of Unix, Has Invented a New Technology that Could Mean Never Having to Buy a CD Again

Will We Live to See Our Brains *Wired* to Gadgets? How About Today?

In the Kingdom of Mao Bell: A Billion Chinese Are Using New Technology to Create the Fastest-Growing Economy on the Planet. But While the Information Wants to Be Free, Do They?

Is National Intelligence an Oxymoron?

Copywrong

Backlash: The Infobahn Is a Big, Fat Joke

The End of Privacy: If Privacy Isn't Already the First Roadkill Along the Information Superhighway, Then it's About to Be

Why the Telcos Are Going to Kick Cable's Butts, and Precisely How the I-Way is Going to Reach Your Home

People Are Supposed to Pay for This Stuff? Crisscrossing the Country, Our Intrepid Correspondent Visits Corporate Labs, Model Living Rooms, and Actual Sofas, to Check Out the Megahyped Interactive Television Prototypes and See Just How Real the 500-Channel, All-Digital, High-Fiber Future Really Is

Viruses Are Good for You: Spawn of the Devil, Computer Viruses May Help Us Realize the Full Potential of the Net

Right Now, There Are No Rules to Keep You from Owning a Bitchin' Corporate Name as Your Own Internet Address

School's Out: The Hyperlearning Revolution Will Replace Public Education

Goodbye, Gutenberg: Pixilating Peer Review Is Revolutionizing Scholarly Journals

The Bit Police: Will the FCC Regulate Licenses to Radiate Bits?

Superhumanism: According to Hans Moravec, by 2040 Robots Will Become as Smart as We Are. And Then They'll Displace Us as the Dominant Form of Life on Earth. But He Isn't Worried—the Robots Will Love Us

Big Brother Wants to Look in Your Bank Account: The U.S. Government Is Constructing a System to Track All Financial Transactions in Real-Time— Ostensibly to Catch Criminals. Does That Leave You With the Warm Fuzzies—or Scare You Out of Your Wits?

Return of the Luddites: A Group of Second-Wave Intellectuals Has Rejected Digital Technology and Declared a Counterrevolution

Where is the Digital Highway Really Heading? The Case for a Jeffersonian Information Policy

IPhone: Will Telephony on the Net Bring the Telcos to Their Knees? Or Will it Allow Them to Take Over the Internet? (And, Oh, Yes, It's Damn Hard to Tap)

Neurobotics: The Future of Computing May Be Gestating—Not in Computer Labs, But in an Obscure Discipline Called Process Control, Where Scientists Have Discovered That a Little Smear of Rat Brain Can Solve One of the Big Problems in Chemical Engineering

America's Futurist Politicians, Al Gore and Newt Gingrich, are Engaged in an Epic Struggle: The Last Time a Battle of This Magnitude Occurred, the New Deal Laid the Foundation of the Modern, Industrial, Bureaucratic State

Wiring Africa: You Go to a Small Village in Africa. People Are Hungry. Is the Real Solution 100 Kilograms of Corn—or an Electronic Mailbox?

The Merry Pranksters Go to Washington: Their "Open Platform" Proposal is the Heart of Gore's Infobahn Policy. They Lead the Coalition Fighting the Clipper Chip. In Short, the Electronic Frontier Foundation Is the Preeminent Defender of Our Civil Rights in Cyberspace. But Just Who Are These People?

Don't Repackage—Redefine! We Have to Resist Media Imperialism—the Tendency to Colonize, to Define New Technologies in Terms of the Old

Cyber-Deterrence: *Wired* Visits the Digital Battlefield of Desert Hammer VI to See Whether the U.S. Army Can Win the Next War Without Firing Another Shot

Mediasaurus: Today's Mass Media Is Tomorrow's Fossil Fuel. Michael Crichton Is Mad as Hell, and He's Not Going to Take it Any More

This Test Is for You: Standardized Testing Is a Communal Rite of Passage. Computer-Adaptive Testing Is About to Make Those Rites Very Individual

Libraries Without Walls for Books Without Pages: Electronic Libraries and the Information Economy

Digital Refusnik: Sven Birkerts Believes that Technology Is Leeching the Spiritual Out of Human Experience

"Anonymous Speech": Imagine Combining Free Speech with Your Right to Privacy

Culture Wars: Francois Mitterrand Has Declared War on Mickey, Madonna, and All-American Culture. Bad News, Francois: Mickey's Winning

Appendix B:
Recording the Data

Compiling the Internet predictions found in this book

The research behind this book was inspired by a suggestion from Lee Rainie, director of the Pew Internet & American Life Project, to Elon University School of Communications faculty members Janna Quitney Anderson and Constance Ledoux Book to assemble a database of predictions made about the Internet during the early 1990s.

In 2001, Book undertook a small preliminary study funded by Pew Internet. Her ensuing 2002 research report, an analysis of several hundred predictive statements found in a brief study of the Lexis-Nexis database, established the following major themes among the statements:

- The Internet will transform society.
- It will transform economies; content will drive the Internet's success.
- The Internet presents security and privacy concerns.
- The Internet's growth is dependent on an efficient and reliable infrastructure.
- The Internet will spawn a new generation of hardware and software.

- The Internet will create a smaller world.
- The Internet will transform America's schools.
- The Internet will impact professions.

Book's report, "Forecasting the Internet: A Retrospective Technology Assessment," was used as the base from which Anderson built an ensuing in-depth study with a completely different methodology aimed at assembling a large, thorough, online database of thousands of predictions made between 1990 and 1995. This more comprehensive study targeted statements made by Internet stakeholders and skeptics, and it included searches through major books of the time, Internet sites, magazines, speeches, research presentations, and newspaper articles. The database is not exhaustive; there are many predictive statements that went unlogged. The collection is a representative sample.

The database:

- Records what the people of the early-to-mid 1990s said about the blooming new communications medium, displays their statements in context, and sorts them into topic and subtopic categories that were determined by the content of the body of predictions obtained.
- Offers a thorough sampling of well-defined and categorized early Internet predictions that can be used as a resource by scholars, historians, students, and researchers. The predictions have been classified by topic and subtopic.
- Identifies and catalogues people—Internet stakeholders and skeptics—who were moved by the Internet explosion of the early 1990s to make some significant prediction(s) about the future of such communications.

The purpose of the work was to scrutinize a great number of the written texts, speeches, and broadcast materials of the early 1990s to identify and document the expectations people began to express. What was everyone saying about the future of this new communications tool? This book puts the most prescient predictions in context in a convenient form for study.

ABOUT THE RESEARCHERS
Anderson is an assistant professor and director of Internet projects at Elon University. Her research falls into the areas of Internet use, Internet history, and online news ethics. She also conducted an earlier study funded by Pew—

"One Neighborhood, One Week on the Internet," a look at how 24 families used the Internet over a one-week span in 2001. She is the co-creator and webmaster for the Elon University/Pew Internet Predictions Database.

Book is an associate professor at Elon University. Her research focuses on digital television and cable policy and regulation.

Other members of the team: Dean Paul Parsons of Elon University's School of Communications lent support in myriad ways. Kate Hickey, director of Elon University's Belk Library, gave the project work space and support. Research librarian Betty Garrison dedicated months of effort to aiding students and finding materials; in addition, she mined and logged many predictions herself. Former Elon faculty member Nadia Watts and 2003 Elon graduate Jennifer Guarino made significant contributions in editing or inputting data. Elon University's Academic Council donated campus office space. School of Communications staff members Phyllis Phillips and Pam Baker assisted with many tasks. The building of the Predictions Database website was accomplished with the help of Alex Lindgren, an Elon University web developer, and Dan Anderson, director of university relations. Harlen Makemson, an assistant professor at Elon, agreed to ask his Media History students to take part in the data-mining effort as part of their course work and helped explain the complicated methodology. Thanks to Makemson, more than 60 Elon University students each spent 18 hours or more seeking, finding, and logging predictive data under the direction of Anderson. Their work constitutes more than a quarter of the found predictions database listings (the students' names are listed below).

Crystal Allen

Patrick Allen

Justen Baskerville

Angela Beckett

Natalie Bizzell

John Bolger

Jason Boone

Lindsay Bradshaw

Erin Bricker

Marian Bruno

Jay Burnham

Lawrence Butler

Lauren Canizaro

Cara Catalfumo

Jason Chick

Kaci Collier

Theresa Cooley

William Culp

Jay Dorn

Kristen Dube

Elizabeth Edwards

Peter Falcone

Peter Fedders

Adam Garber

Barbara Goodrich
Mathea Gulbranson
Nichelle Harrison
Abbey Heiskell
Shavanna Jagrup
Kathleen Johnson
David Kafoure
Tiffany Kildale
Kelly Kohlhagen
Anne Komorowski
Kevin Krout
Ellie Lightburn
Brandi Little
Travis Lusk
Casey Marge
Rory McAlister
Jennifer Meyer
Erin Moseley
Diana Nolan
Shawna Pagano

Bradley Pinkerton
Kristin Ries
Melanie Sampson
Nicholas Schmidt
Carrie Scott
Tim Severs
Ian Smith
Barry Smoot
Shawn Stevens
Ben Stewart
Larry Stotler
Amanda Strickland
Matt Sturmfelz
Kellen Taylor
Elizabeth Tencer
Evelyn Uhlfelder
Amanda Vellucci
Abigail Wahl
Meghan Walsh
Laura Wright

DATA ABOUT THE ELON UNIVERSITY/PEW INTERNET PREDICTIONS DATABASE

Sources from which Predictions Were Mined and Logged: Thousands of websites (including writings, speeches, and video and audio transcripts) and tens of thousands of periodical articles were searched. Nearly 40 books—all included in this book's bibliography—were thoroughly searched, and many additional books were assessed and found not to include predictions.

Number of Files Entered and Edited: 4,217 (1,475 found by Elon University students, others by project director, staff assistants)

Number of Predictors: About 1,000

Number of Different Topic Fields into Which Predictions Were Sorted: Eight, including the following:

Community/Culture: 485 predictions
Global Relationships/Politics: 310 predictions
Economic Structures: 406 predictions

Information Infrastructure: 1,004 predictions
Communication: 337 predictions
General, Overarching Remarks: 228
Getting, Sharing Information: 971 predictions
Controversial Issues: 477 predictions
See table B.1 for a presentation of different subtopic fields into which predictions were sorted.

Table B.1.   Subtopic Fields (68) into which Predictions Were Sorted

Under Community/Culture	Under Global Relationships/Politics	Under Economic Structures
Cyberpunks/Hackers: 8;	Campaigns/Voting: 21;	E-cash: 41;
Ethics/Values: 53;	Creating a Smaller World: 16;	E-commerce: 98;
Human-Machine Interaction: 66;	Democracy: 116;	Employment: 31;
Information Overload: 22;	Government: 29;	Gambling: 19;
MOOs/MUDs/B-Boards/	Peacekeeping/Warfare: 37;	Microtransactions: 8;
Newsgroups: 27;	Third-World Nations: 7;	Shopping: 44;
Relationships: 46;	General: 84	Telecommuting: 26;
Social Withdrawal/Addiction: 21;		Tax Issues: 7
Virtual Communities: 91;		
General: 149		

Under Information Infrastructure	Under Communication	Under Getting, Sharing Information
Bandwidth: 50;	E-mail: 69;	Advertising/PR: 51;
Cost/Pricing: 91;	Internet Telephony: 20;	Crisis Management: 6;
Internet Appliances: 92;	Security/Encryption: 171;	Libraries/Databases: 67;
Internet Service Providers: 26;	Video Conferencing: 8;	E-learning: 222;
Language/Interface/Software: 122;	Viruses/Worms: 12;	Gaming: 7;
Number of Users: 20;	General: 57	Intelligent Agents/AI: 65;
Open Access: 44;		Journalism/Media: 63;
Pipeline/Switching/Hardware: 117;		Medical/Professional: 58;
Protocols: 42;		Music: 17;
Role of Government/Industry: 109;		Newspapers: 39;
Universal Service: 24;		Publishing: 68;
Wireless Technologies: 38;		TV/Films /Video: 38;
General: 229		Virtual Reality: 27;
		General: 243

Under Controversial Issues
Anonymity/Personal Identity: 16;
Censorship/Free Speech: 62;
Copyright/Intellectual
Property/Plagiarism: 107;
Crime/Fraud/Terrorism: 38;
Defamation/Libel: 5;
Digital Divide: 78;
Jurisdiction/Control: 41;
Privacy/Surveillance: 89;
Pornography: 21;
General: 20

Suggested Readings

Readings for more on the history of the Internet

True Names (1981) by **Vernor Vinge**. Three years before William Gibson came out with the influential pre-Internet book *Neuromancer*, Vinge wrote a fictional novella that closely reflected the online reality to come. In an April 5, 1999, interview with Andrew Leonard of *Salon*, Vinge said: "When I was first writing science fiction in the early '60s, it was easy to have ideas that it turned out didn't happen for 20 or 30 years. But now it is very hard to keep up—in part because the people who are making things happen have absorbed science fiction's mind-set of scenario building and technological brainstorming. They are driving ahead of their headlights too, now, and things are going very, very fast." The *True Names* story line includes group use of a common networked communication similar to the MUDs (multi-user domains) that were to follow in the real world; cryptography is utilized to provide security online; and the users he portrays are Internet addicts.

Neuromancer (1984) by **William Gibson.** This novel was ballyhooed as the first work of cyberpunk fiction, and it spurred a great deal of study, interest, and imitators. The cyberpunk genre is characterized by a depiction of a dark future in which pollution, wars, economics, and/or the rise of technology force people to struggle for survival. Some critics say that Bruner's *Shockwave Rider* and Vinge's *True Names* gave an outline of cyberspace, but Gibson gave it a name: Gibson's *Neuromancer* introduced the formal title "cyberspace" and described virtual reality. "Cyberspace," Gibson wrote. "A consensual hallucination experienced daily by billions of legitimate operators, in every nation, by children being taught mathematical concepts. . . . A graphical representation of data abstracted from the banks of every computer in the human system. Unthinkable complexity. Lines of light ranged in the non-space of the mind, clusters and constellations of data. Like city lights, receding."

Technopoly: The Surrender of Culture to Technology (1993) by **Neil Postman**. The cultural critic and professor of media ecology presents his view of a future that features "progress without limits," "rights without responsibilities," "technology without cost" and a "moral center" replaced by "efficiency, interest and economic advance"—what Postman labels a "Technopoly"—the prime example of which he says is the United States. The elements of truth in his arguments are valuable; an understanding of the negatives can help soften the blows certain to be delivered along with the positives of new technologies.

The Virtual Community (1993) by **Howard Rheingold**. A pioneering illuminator of the Internet culture, Rheingold discusses the benefits and possible problems for society brought forth by the popularization of networked groups of people. In the introduction, he says: "I have written this book to help inform a wider population about the potential importance of cyberspace to political liberties. . . . Although I am enthusiastic about the liberating potentials of computer-mediated communications, I try to keep my eyes open for the pitfalls of mixing technology and human relationships." Available online at http://www.well.com/user/hlr/vcbook/vcbookintro.html

The Gutenberg Elegies: The Fate of Reading in an Electronic Age (1993) by **Sven Birkerts**. Birkerts shares a collection of his essays through which he conducts what he calls an "inquiry into the place of reading and sensibility in what is becoming an electronic culture." He says "technological ingenuity will set the agenda" and "Americans will follow," labeling this an "argument between technology and the soul." He urges readers to face down the digital future and "refuse it."

City of Bits (1994) by **William J. Mitchell**. Written by the dean of the School of Architecture and Planning at the Massachusetts Institute of Technology. The traditional printing of this book was followed in the spring and summer of 1995 with a companion online issue—what was labeled as "the first full-text interactive book on the World Wide Web." It is one of the finest looks at what may be that was generated in this era—equal in imagination and forward-thinking to Nicholas Negroponte's better-publicized *Being Digital*, and just as cleverly rendered. http://mitpress2.mit.edu/e-books/City_of_Bits/

Life After Television (1994) by **George Gilder.** Known as one of the best-respected high-tech futurist/consultants of the 1990s, the prolific Gilder shared his vision in his books and magazine articles. Some of his classic work from the early 1990s can be read on his website: http://www.gildertech.com/public/telecosm_series/lifetv.html

The Right to Privacy (1994) by **Ellen Alderman** and **Caroline Kennedy.** The word "privacy" does not appear in the United States Constitution, these authors point out. Yet people consider it a right, and it is a right that is being eroded thanks to technology. Alderman and Kennedy discuss many new issues of the time, including companies' monitoring of employees' e-mail correspondence.

Out of Control: The New Biology of Machines, Social Systems and the Economic World, (1994) by **Kevin Kelly.** Former *Wired* magazine editor Kelly writes about systems and networks, saying that we will make our machines biological in order to manage them, adding that we are connecting everything to everything, and our entire culture is migrating to a "network culture" and a new network economics. Kelly is a co-founder of Long Bets, an organization that was founded to improve long-term thinking. To read predictions made since 2002 about the future, see the site http://www.longbets.org/

Being Digital (1995) by **Nicholas Negroponte.** One of the high-visibility ambassadors of the Internet in the 1990s, Negroponte wrote and spoke in glowing terms of "being digital," seeing a glowing future for the world. The co-founder of MIT's Media Lab offered here an introduction to the possibilities of digital communication for the uninitiated. He had helped bankroll the start-up of *Wired* magazine in 1993, and his monthly column for that publication—considered the *Rolling Stone* of the technology age—forms the basis for this book, considered to be a classic predictive book about the potential of networking. http://archives.obs-us.com/obs/english/books/nn/bdcont.htm

Life on the Screen (1995) by **Sherry Turkle.** This fascinating book takes a look at the changing impact of the computer as it went from a business machine in the 1980s to a personal communications and information device in the 1990s. Turkle, also of MIT, looks at the activity in multi-user domains and research being done with intelligent agents at MIT, and she writes about the evolution of the way humans see themselves and machines, explaining that multiple identities may emerge.

Silicon Snake Oil: Second Thoughts on the Information Highway (1995) by **Clifford Stoll**. Computer user and network community member Stoll describes what he sees as the potential ramifications of embracing the digital age thoughtlessly and in an overenthusiastic manner. He decries the hyping of the info highway, saying "every hour that you're behind the keyboard is sixty minutes that you're not doing something else."

The Electronic Republic (1995) by **Lawrence Grossman**. In this book, the former president of NBC News and head of PBS takes a critical look at the pros and cons of the potential new roles for media and networked communications in a democracy. He points out the likelihood that a "direct democracy" is growing due to instant communications and sees that the general citizenry is the "new fourth branch of government." He proposes that U.S. citizens and leaders open their eyes to the changes and make adjustments for a digital future.

The Road Ahead (1995) by **Bill Gates** with **Nathan Myhrvold** and **Peter Rinearson**. The chairman of software giant Microsoft and two of his associates share the thoughts of their corporation regarding the future of networked communications. The book, written in Bill's "voice," was published as an explanatory guide to get the general public ready to buy into more cool technology products. It weaves together all of the common wisdom of the previous two decades regarding what the future would likely bring.

Creating a New Civilization: The Politics of the Third Wave (1995) by **Alvin** and **Heidi Toffler**. The Tofflers say that just as the agricultural first wave has given way to the second wave industrial age, it in turn has yielded to the third wave—the knowledge revolution. They outline the differences and prescribe the need for change, saying the third wave is breaking down the old bureaucracies and creating the need for new approaches to government, business, and other social structures. They write: "The third wave is not just a matter of technology and economics. It involves morality, culture and ideas as well as institutions and political structure. In short, a true transformation in human affairs."

The Second Media Age (1995) by **Mark Poster**. This critical evaluation of the concepts of media and technology examines theories of postmodernity in relation to the new media and the debate over multiculturalism. Poster argues that new developments in electronic media, such as the Internet and virtual reality, may so alter our habits of communication and so deeply reposition our identities that the designation "a second media age" is justified. He discusses the contributions of theorists, including Baudrillard, Lyotard, Habermas, Haraway, and Guattari, and builds on his previous book, *The Mode of Information.*

Rebels Against the Future: The Luddites and Their War on the Industrial Revolution (1995) by **Kirkpatrick Sale.** The main point of this book is to describe the uprising led by some of the first victims of the industrial revolution. In 1811 and 1812, cottage workers in the textile trades fought an armed rebellion against the moneyed interests engineering change by building centralized, mechanized factories. Sale concludes by drawing parallels to the incursion of digital and networked technologies, siding with neo-Luddites.

Netizens: On the History and Impact of Usenet and the Internet (1997) by **Ronda** and **Michael Hauben.** A netizen is a person who has exhibited a commitment to model citizenship in an online community. This history of the Internet is told through netizens' eyes, mostly dwelling on the participants in Usenet, the vast array of bulletin board–like message areas where people found discussions about everything from the mundane daily life to scientific minutia to pop culture.

Where Wizards Stay Up Late: The Origins of the Internet (1998) by **Katie Hafner** and **Matthew Lyon.** This extensively researched look at the beginning of the Internet is considered a classic. The authors did hundreds of interviews and dug through thousands of pages of reports and other archived materials. *Wizards* delves—in layman's language—into the fine details, strange stories, successes, and failures of the new communications medium. Accomplished tech writer Hafner is also the co-author of *Cyberpunk: Outlaws and Hackers on the Computer Frontier* (1995) and the author of *The Well: A Story of Love, Death and Real Life in the Seminal Online Community.*

Weaving the Web: The Original Design and Ultimate Destiny of the World Wide Web by its Inventor (1999) by **Tim Berners-Lee.** The author of HTML tells his version of the story of the origins of the Web. Berners-Lee studied physics at Oxford before going to work at CERN (a particle physics laboratory near Geneva), where he decided to program his new computer so that he would create "a space in which anything could be linked to anything." The book covers major issues such as security, privacy, and free speech, but it also tells the inside story of the origins of such things as the "URL" (he favored "URI"), how "http://" became the prefix of choice, and how ".com" and other domains came to be.

Architects of the Web: 1,000 Days That Built the Future of Business (1999) by **Robert H. Reid**. This is a study of the start of the Internet from a business perspective. Each chapter tells the story of a different Web entrepreneur, including Mark Pesce of VRML, Marc Andreessen of Mosaic, Halsy Minor of CNET, and Jerry Yang of Yahoo!

Inventing the Internet: Inside Technology (2000) by **Janet Abbate**. A look at the forty years of behind-the-scenes work that went into making the Internet what it is today, from David Baran's work at RAND through Tim Berners-Lee's breakthroughs at CERN and beyond.

The Internet Galaxy (2001) by **Manuel Castels**. An information-age analyst looks at how the origins of the Internet now influence our cultural and economic structures. This book also includes a brief but interesting look at the development of the Internet from the 1960s to the 1990s.

Wired: A Romance (2003) by **Gary Wolf**. An insider's look (Wolf was a *Wired* writer and editor) at the founding of the highly influential *Wired* magazine.

HISTORY OF THE INTERNET
For online information about the history of the Internet, try these links:

The Internet Society: http://www.isoc.org

Hobbes's Internet Timeline: http://www.zakon.org/robert/internet/timeline/#1990s

The *Wired* magazine archive: http://www.wired.com/wired/archive/

INFORMATION OVERLOAD
Sites with information about information overload:

Neil Postman's classic 1990 speech for the German Informatics Society, "Informing Ourselves to Death": http://www.eff.org/Net_culture/Criticisms/informing_ourselves_to_death.paper

David Kirsh, of the department of cognitive science, University of California–San Diego, a 2000 essay for *Intellectica* titled "A Few Thoughts on Cognitive Overload": http://icl-server.ucsd.edu/~kirsh/Articles/Overload/published.html

Nikolai Bezroukov, "Information/Word Overload Annotated Webliography"—a 1999 compendium of information linking to various sources on the topic: http://www.softpanorama.org/Social/overload.shtml

Mark R. Nelson's early 1990s research paper, "We Have the Information You Want, But Getting It Will Cost You: Being Held Hostage By Information Overload," carried on the ACM site: http://www.acm.org/crossroads/xrds1-1/mnelson.html

Peter Lyman and Hal R. Varian, "How Much Information? 2003": http://www.sims.berkeley.edu/research/projects/how-much-info-2003/

F. Heylighen, "Change and Information Overload," a 1999 article carried on the Principia Cybernetica site: http://pespmc1.vub.ac.be/CHINNEG.html

D. M. Griffiths, a 2004 guide to helping people make information-management decisions: http://www.managing-information.org.uk/introduction.htm

Paul Waddington, 1994 research paper, "Dying for Information: A Report on the Effects of Information Overload in the UK and Worldwide": http://www.cni.org/regconfs/1997/ukoln-content/repor~13.html

GAIA, NETWORKS, AND DYNAMIC SYSTEMS

To read more about Gaia, networks, and dynamic systems, see:

Ashby, Ross. *Introduction to Cybernetics.* New York: Wiley, 1956.

Barabasi, Albert-Laszlo. *Linked: The New Science of Networks.* New York: Perseus Publishing, 2002.

Bateson, Gregory. *Mind and Nature: A Necessary Unity.* New York: Dutton, 1979.

von Bertalanffy, Ludwig. *General System Theory.* New York: Braziller, 1968.

Buchanan, Mark. *Nexus: Small Worlds and the Groundbreaking Science of Networks.* New York: W. W. Norton, 2002.

Capra, Fritjof. *The Web of Life.* New York: Random House, 1996.

Dawkins, Richard. *The Selfish Gene.* New York: Oxford University Press, 1976.

de Sola Pool, Ithiel. *Technologies of Freedom: On Free Speech in an Electronic Age.* Cambridge: Harvard University Press, 1983.

Etheredge, Lloyd S. *Politics in Wired Nations: Selected Writings of Ithiel de Sola Pool.* New Brunswick, NJ: Transaction Publishers, 1998.

Kochen, Manfred, ed. *The Small World: A Volume of Recent Research Commemorating Ithiel de Sola Pool, Stanley Milgram, and Theodore Newcomb.* Norwood, NJ: Ablex Publishing Company, 1989.

Lilienfeld, Robert. *The Rise of Systems Theory.* New York: Wiley, 1978.

Lovelock, James. *Gaia.* New York: Oxford University Press, 1979.

———. *Healing Gaia.* New York: Harmony Books, 1991.

Maturana, Humberto, and Francisco Varela. *Autopoiesis and Cognition.* Dordrecht, Holland: De. Reidel, 1980.

Schneider, Stephen, and Penelope Boston, eds. *Scientists on Gaia.* Cambridge: MIT Press, 1991.

Taylor, Mark. *The Moment of Complexity: Emerging Network Culture.* Chicago: University of Chicago Press, 2001.

Watts, Duncan J. *Six Degrees: The Science of a Connected Age.* New York: W. W. Norton, 2003.

Weiner, Norbert. *Cybernetics.* Cambridge: MIT Press, 1961.

HISTORY OF THE TELEGRAPH

To read more about the history of the telegraph, see:

Blondheim, Menahem. *News Over the Wires.* Cambridge: Harvard University Press, 1994.

Briggs, Charles F., and Augustus Maverick. *The Story of the Telegraph.* New York: Rudd and Carleton, 1863.

Coe, Lewis. *The Telegraph: A History of Morse's Invention and Its Predecessors in the United States.* Jefferson, NC: McFarland and Company, 1993.

Oslin, George P. *The Story of Telecommunications.* Macon, GA: Mercer University Press, 1992.

Parsons, Frank. *The Telegraph Monopoly.* Philadelphia: C. F. Taylor, 1899.

Rosewater, Victor. *History of Cooperative News-Gathering in the United States.* New York: D. Appleton and Company, 1930.

Standage, Tom. *The Victorian Internet*. New York: Walker and Company, 1998.

Thompson, Robert Luther. *Wiring a Continent: The History of the Telegraph Industry in the United States, 1832–1866*. Princeton, NJ: Princeton University Press, 1947.

Williams, Frederick. *The Communications Revolution*. Beverly Hills, CA: Sage, 1982.

HISTORY OF RADIO
To read more about the history of radio, see:

Aitken, Hugh G. J. *Syntony and Spark: The Origins of Radio*. New York: Wiley, 1976.

———. *The Continuous Wave: Technology and American Radio, 1900–1932*. Princeton, NJ: Princeton University Press, 1985.

Baker, W. J. *A History of the Marconi Company*. London: Methuen and Co. Ltd., 1970.

Barnouw, Eric. *A Tower in Babel: A History of Broadcasting in the United States to 1933*. New York: Oxford University Press, 1966.

Bilby, Kenneth. *The General: David Sarnoff and the Rise of the Communications Industry*. New York: Harper and Row, 1986.

Douglas, Susan J. *Inventing American Broadcasting, 1899–1922*. Baltimore: Johns Hopkins University Press, 1987.

Jolly, W. P. *Marconi*. London: Constable and Company, 1972.

Lewis, Tom. *Empire of the Air: The Men Who Made Radio*. New York: HarperCollins, 1991.

Reith, J. C. W. *Broadcast Over Britain*. London: Hodder and Stoughton, 1924.

Schubert, Paul. *The Electric Word*. New York: The Macmillan Company, 1928.

HISTORY OF THE TELEPHONE
To read more about the history of the telephone, see:

Boettinger, H. M. *The Telephone Book: Bell, Watson, Vail and American Life, 1876–1983*. New York: Stearn, 1983.

Bruce, Robert V. *Bell: Alexander Graham Bell and the Conquest of Solitude*. Boston: Little, Brown, 1973.

Coe, Lewis. *The Telephone and Its Several Inventors: A History*. Jefferson, NC: McFarland, 1995.

de Sola Pool, Ithiel. *Forecasting the Telephone: A Retrospective Technology Assessment of the Telephone*. Norwood, NJ: Ablex Publishing Company, 1983.

Fischer, Claude S. *America Calling*. Berkeley: University of California Press, 1992.

Grosvenor, Edwin S., and Morgan Wesson. *Alexander Graham Bell: The Life and Times of the Man Who Invented the Telephone*. New York: Harry Abrams, 1997.

Kingsbury, John E. *The Telephone and Telephone Exchanges: Their Invention and Development*. Longmans, Green, 1915; reprinted by Arno Press, 1972.

Prescott, George B. *Bell's Electric Speaking Telephone: Its Invention, Construction, Application, Modification and History*. New York: D. Appleton, 1884; reprinted by Arno Press, 1972.

Shiers, George. *The Telephone: An Historical Anthology*. New York: Arno Press, part of the series *Historical Studies in Telecommunications*, 1977.

HISTORY OF TELEVISION

To read more about the history of television, see:

Abramson, Albert. *History of Television, 1880–1941*. Jefferson, NC: McFarland, 1987.

Barnouw, Erik. *Tube of Plenty: The Evolution of American Television*. New York: Oxford, 1990.

Burns, R. W. *John Logie Baird: Television Pioneer*. London: Institution of Electrical Engineers, 2000.

Fink, D. G. "Television Broadcasting in the United States, 1927–1950." *Proceedings of the IRE*. February 1951, 116–23.

Fisher, David E. and Marshall J. *Tube: The Invention of Television*. Washington, DC: Counterpoint, 1996.

Godfrey, Donald. *Philo T. Farnsworth: Father of Television*. Salt Lake City: University of Utah Press, 2001.

Hutchinson, Thomas H. *Here Is Television: Your Window to the World*. New York: Hastings House, 1946.

Jensen, Axel G. "The Evolution of Modern Television." *Journal of the Society of Motion Picture and Television Engineers*. November 1954, 174–88.

Shiers, George. "Historical Notes on Television Before 1900." *Journal of the Society of Motion Picture and Television Engineers*. March 1977, 129–37.

Slotten, Hugh R. *Radio and Television Regulation: Broadcast Technology in the United States, 1920–1960.* Baltimore: Johns Hopkins University Press, 2000.

Udelson, Joseph H. *The Great Television Race: History of the American Television Industry, 1925–1941.* University of Alabama Press, 1982.

Yanczer, Peter F. *The Mechanics of Television: The Story of Mechanical Television.* St. Louis: the author, 1987.

Bibliography

Detailed citations of individual direct quotations from the span of 1990 to 1995 found throughout this book are available online at the Elon University/Pew Internet database: www.elon.edu/predictions/advanced.aspx. One of the easiest ways to use the search engine to find a specific citation for a specific quote is to isolate a segment of that quote that is likely to be unique (for instance, "the Internet is a network of networks, no one group or person is in charge") and copy and paste it into the "Search" box of the Early 1990s database, enclosing the segment in quote marks. This will take you to a page of "hits" for that search. Click on a blue headline here and the search engine shows the complete entry page tied to that headline. The full citation detail is found at the *bottom* of each quote's page.

The following are other major resources used as background in writing the supporting narrative aspects of this book.

Abbate, Janet. *Inventing the Internet.* Cambridge: MIT Press, 1999.

Abramson, Albert. *History of Television, 1880–1941.* Jefferson, NC: McFarland, 1987.

———. *Zworykin: Pioneer of Television.* Urbana: University of Illinois Press, 1995.

Agre, Philip E., and Marc Rotenberg, eds. *Technology and Privacy: The New Landscape.* Cambridge: MIT Press, 1997.

Aitken, Hugh, and G. J. Syntony. *Spark: The Origins of Radio.* New York: Wiley, 1976.

———. *The Continuous Wave: Technology and American Radio, 1900–1932.* Princeton, NJ: Princeton University Press, 1985.

Alderman, Ellen, and Caroline Kennedy. *The Right to Privacy.* New York: Knopf, 1994.

Archer, Gleason L. *History of Radio to 1926.* New York: American Historical Society, 1938.

Arvin, W. B. "See with Your Radio." *Radio News,* September 1925, 384–87.

Ashby, Ross. *Introduction to Cybernetics.* New York: Wiley, 1956.

Asimov, Isaac. *I, Robot.* New York: Gnome Press, 1950.

Baker, W. J. A. *History of the Marconi Company.* London: Methuen and Co. Ltd., 1970.

Barabasi, Albert-Laszlo. *Linked: The New Science of Networks.* Cambridge: Perseus Publishing, 2002.

Baran, Paul, and Sharla P. Boehm. "On Distributed Communications: II Digital Simulation of Hot Potato Routing in a Broadband Distributed Communications Network." Memorandum RM-3103-PR August 1964 Available [Online]: http://www.rand.org/publications/RM/RM3103/ [01 July 2004].

Barnouw, Erik. *A Tower in Babel: A History of Broadcasting in the United States to 1933.* New York: Oxford University Press, 1966.

———. *Tube of Plenty: The Evolution of American Television.* New York: Oxford University Press, 1990.

Bateson, Gregory. *Mind and Nature: A Necessary Unity.* New York: Dutton, 1979.

Bell, A.G. "Selenium and the Photophone." *Nature* 21 (1880), 22, 23.

———. *The Bell Telephone: The Deposition of Alexander Graham Bell in the Suit Brought by the United States to Annul the Bell Patents.* American Bell Telephone, 1908; reprinted by Arno Press, 1974.

Berners-Lee, Tim, with Mark Frischetti. *Weaving the Web: The Original Design and Ultimate Destiny of the World Wide Web.* San Francisco: HarperCollins, 1999.

von Bertalanffy, Ludwig. *General System Theory.* New York: Braziller, 1968.

Bezroukov, Nikolai. "Information/Word Overload Annotated Webliography. —a 1999 compendium." Available [Online] http://www.softpanorama.org/Social/ overload.shtml. [9 August 2004].

Bidwell, G. L. "Television Arrives." *QST* July 1925, 9–14.

Bilby, Kenneth. *The General: David Sarnoff and the Rise of the Communications Industry.* New York: Harper and Row, 1986.

Birkerts, Sven. *The Gutenberg Elegies: The Fate of Reading in an Electronic Age.* Winchester, MA: Faber & Faber, 1993.

Blondheim, Menahem. *News Over the Wires.* Cambridge: Harvard University Press, 1994.

Boettinger, H. M. *The Telephone Book: Bell, Watson, Vail and American Life, 1876–1983.* New York: Stearn, 1983.

Bois, Daniel. "The 1980s and 1990s Microelectronics Logbook: Guidelines for the Future." Research paper reprinted in *From Microelectronics to Nanoelectronics: Roadmaps and Challenges.* New York: Wiley, 1999.

Boyle, James. *Shamans, Software, and Spleens: Law and the Construction of the Information Society.* Cambridge: Harvard University Press, 1996.

Braudel, Fernand. *The Structures of Everyday Life: The Limits of the Possible— Civilization and Capitalism, 15th–18th Century.* Berkeley, CA: University of California Press, 1992.

Briggs, Charles F., and Augustus Maverick. *The Story of the Telegraph.* New York: Rudd and Carleton, 1863.

Brockman, John. *Digerati: Encounters with the Cyber Elite.* Emeryville, CA: Publishers Group West, 1996.

———, ed. *The Next Fifty Years: Science in the First Half of the Twenty-First Century.* New York: Vintage, 2002.

Brooks, Rodney A. *Flesh and Machines: How Robots Will Change Us.* New York: Pantheon, 2002.

Brooks, Rodney A., and A. M. Flynn. "Fast, Cheap, and Out of Control: A Robot Invasion of the Solar System." *Journal of the British Interplanetary Society*, October 1989, 478–85.

Bruce, Robert V. *Bell: Alexander Graham Bell and the Conquest of Solitude.* Boston: Little, Brown, 1973.

Buchanan, Mark. *Nexus: Small Worlds and the Groundbreaking Science of Networks.* New York: W.W. Norton, 2002.

Burns, R. W. *John Logie Baird: Television Pioneer.* London: Institution of Electrical Engineers, 2000.

Burstein, Daniel, and David Kline. *Road Warriors: Dreams and Nightmares Along the Information Highway.* New York: Dutton, 1995.

Bush, Vannevar. "As We May Think." *The Atlantic Monthly* 176, 1 (July 1945), 101–108.

Capra, Fritjof. *The Web of Life.* New York: Random House, 1996.

Carey, G. R. "Commenting on 'Seeing by Electricity.'" *Scientific American,* June 1880.

Carneal, Georgette. *A Conqueror of Space: An Authorized Biography of the Life and Work of Lee De Forest.* New York: H. Liveright, 1930.

Castels, Manuel. *The Internet Galaxy.* Oxford: Oxford University Press, 2001.

Chapuis, Robert J. *100 Years of Telephone Switching 1878–1978.* Part 1: *Manual and Electromechanical Switching, 1878–1960s.* Amsterdam: North-Holland Publishing Studies in Telecommunications, 1982.

Church, A. "Recent Developments in Television." *Nature,* September 30, 1933, 502–505.

Coe, Lewis. *The Telegraph: A History of Morse's Invention and Its Predecessors in the United States.* Jefferson, NC: McFarland, 1993.

———. *The Telephone and Its Several Inventors: A History.* Jefferson, NC: McFarland, 1995.

Cohen, Frederick B. *Protection and Security on the Information Superhighway.* New York: Wiley, 1995.

Computer Museum History Center. "Internet History 1962–1992." Available [Online]: http://www.computerhistory.org/exhibits/internet_history/index.page [9 July 2004].

CSIS (Center for Strategic and International Studies). Panel Reports 0899-0352. "Reinventing Diplomacy in the Information Age: A Report of the CSIS Advisory Panel on Diplomacy in the Information Age." Washington, D.C.: Center for Strategic and International Studies, 1998.

Czitrom, Daniel J. *Media and the American Mind: From Morse to McLuhan*. Chapel Hill: University of North Carolina Press, 1982.

Dawkins, Richard. *The Selfish Gene*. Oxford: Oxford University Press, 1976.

De Forest, Lee. *Father of Radio: The Autobiography of Lee De Forest*. Chicago: Wilcox and Follett, 1950.

de Sola Pool, Ithiel. *Technologies of Freedom: On Free Speech in an Electronic Age*. Cambridge: Harvard University Press, 1983.

———. *Forecasting the Telephone: A Retrospective Technology Assessment of the Telephone*. Norwood, NJ: Ablex Publishing Company, 1983.

Dinsdale, A. A. *First Principles of Television*. New York: Wiley, 1932, reprinted, Arno Press, 1971.

Douglas, Susan J. *Inventing American Broadcasting, 1899–1922*. Baltimore: Johns Hopkins University Press, 1987.

Drake, William. "Policies for the National and Global Information Infrastructures." In William Drake, ed., *The New Information Infrastructure: Strategies for U.S. Policy*. New York: Century Foundation Press, 1995, 345–78.

DuBoff, Richard B. "Business Demand and the Development of the Telegraph in the United States, 1844–1860." *Business History Review*, Winter 1980, 459–79.

———. "The Telegraph in Nineteenth-Century America: Technology and Monopoly." *Comparative Studies in Society and History* 26 (1984), 571–86.

Dunlap, Orrin E., Jr. *Marconi: The Man and His Wireless*. New York: Macmillan, 1937.

Dutton, William. *Society on the Line: Information Politics in the Digital Age*. Oxford: Oxford University Press, 1999.

Eckhardt, George H. *Electronic Television*. Chicago: Goodheart-Willcox Co., 1936, reprinted, Arno Press, 1974.

Elon University/Pew Internet and American Life Project Predictions Database. "Imagining the Internet." Available [Online]: http://www.elon.edu/predictions [9 August 2004].

Etheredge, Lloyd S. *Politics in Wired Nations: Selected Writings of Ithiel de Sola Pool*. New Brunswick, NJ: Transaction Publishers, 1998.

Field, Henry M. *History of the Atlantic Telegraph*. New York: Charles Scribner and Co., 1866.

Fink, D.G. "Television Broadcasting in the United States, 1927–1950." Proceedings of the IRE, February 1951, 116–23.

Fischer, Claude S. *America Calling: A Social History of the Telephone to 1940.* Berkeley: University of California Press, 1992.

Fisher, David E. and Marshall J. *Tube: The Invention of Television.* Washington, D.C.: Counterpoint, 1996.

FRC/FCC. "The Evolution of Television: 1927–1943. —as reported in the Annual Reports of the Federal Radio Commission and the Federal Communications Commission." *Journal of Broadcasting,* Summer 1960, 199–240.

Gates, Bill, Nathan Myhrvold, and Peter Rinearson. *The Road Ahead.* New York: Penguin USA, 1995.

Gelernter, David. *The Muse in the Machine: Computerizing the Poetry of Human Thought.* New York: The Free Press, 1994.

Gergen, Kenneth J. *The Saturated Self: Dilemmas of Identity in Contemporary Life.* New York: BasicBooks, 1991.

Gibson, William. *Neuromancer.* New York: Ace Publishing, 1984.

Gilder, George. *Life After Television.* New York: Norton, 1994.

Godfrey, Donald G. *Philo T. Farnsworth: The Father of Television.* Salt Lake City: University of Utah Press, 2001.

Granovetter, Mark. "The Strength of Weak Ties." *American Journal of Sociology* 78, 1973, 1360–80.

Grossman, Lawrence. *The Electronic Republic.* New York: Viking, 1995.

Grosvenor, Edwin S., and Wesson Morgan. *Alexander Graham Bell: The Life and Times of the Man Who Invented the Telephone.* New York: Harry Abrams, 1997.

Habermas, Jurgen. *The Theory of Communicative Action.* Boston: Beacon Press, 1987.

Hafner, Katie, and Matthew Lyon. *Where Wizards Stay Up Late: The Origins of the Internet.* New York: Simon and Schuster, 1996.

Hafner, Katie, and John Markoff. *Cyberpunks: Outlaws and Hackers in the Computer Frontier.* New York: Touchstone, 1995.

Hardy, Ian R. The Evolution of ARPANET Email. Ph.D. diss., University of California at Berkeley 1996. Available [Online]: http://www.ifla.org/documents/internet/hari1.txt [9 July 2004].

Harlow, Alvin F. *Old Wires and New Waves*. New York: D. Appleton-Century Company, 1936.

Hauben, Michael and Ronda. *Netizens: On the History and Impact of Usenet and the Internet*. Los Alamitos, CA: IEEE Computer Society Press, 1997.

Heylighen, F. "Change and Information Overload." A 1999 article on the Principia Cybernetica site. Available [Online] http://pespmc1.vub.ac.be/CHINNEG.html. [9 August 2004].

Hiltz, Starr Roxanne, and Murray Turoff. *Network Nation*. Cambridge: MIT Press, 1995.

Hutchinson, Thomas H. *Here Is Television: Your Window to the World*. New York: Hastings House, 1946.

Inglis, Andrew F. *Behind the Tube: A History of Broadcasting Technology and Business*. Stoneham, MA: Focal Press, 1990.

Internet Society ISOC-LA. Internet 30th Anniversary-UCLA-1999-09-02. Available [Online]: http://www.isoc-la.org/30Anniv-UCLA.htm [9 July 2004].

Jenkins, Charles Francis. *Radiomovies, Radiovision, Television*. Washington, D.C.: National Capital Press, Inc., 1929.

Jensen, Axel G. "The Evolution of Modern Television." *Journal of the Society of Motion Picture and Television Engineers*, November 1954, 174–88.

Jolly, W. P. *Marconi*. London: Constable and Company, 1972.

Jones, Steve, ed. *Virtual Culture*. London: Sage, 1997.

Kahin, Brian, and Charles R. Nesson. *Borders in Cyberspace: Information Policy and the Global Information Infrastructure*. Cambridge: MIT Press, 1997.

Kehoe, Brendan P. *Zen and the Art of the Internet: A Beginner's Guide*. Saddle River, NJ: Prentice Hall, 1996.

Kelly, Kevin. *Out of Control: The New Biology of Machines, Social Systems, and the Economic World*. Reading, MA: Addison Wesley, 1994.

Kingsbury, John E. *The Telephone and Telephone Exchanges: Their Invention and Development*. New York: Longmans, Green, 1915; reprinted by Arno Press, 1972.

Kirsh, David. "A Few Thoughts on Cognitive Overload." 2000 essay for *Intellectica* magazine, Available [Online]: http://icl-server.ucsd.edu/~kirsh/Articles/Overload/published.html [9 August 2004].

Kirshon, John W. *Chronicle of the 20th Century.* London: Dorling Kindersley, 1995.

Kochen, Manfred, ed. *The Small World: A Volume of Recent Research Commemorating Ithiel de Sola Pool, Stanley Milgram, and Theodore Newcomb.* Norwood, NJ: Ablex Publishing Company, 1989.

Kurzweil, Ray. *The Age of Spiritual Machines.* New York: Viking, 1999.

Leiner, Barry M., Vinton G. Cerf, David D. Clark, Robert E. Kahn, Leonard Kleinrock, Daniel C. Lynch, Jon Postel, Larry G. Roberts, and Stephen Wolff. "A Brief History of the Internet." Available [Online]: http://www.isoc.org/internet-history/brief.html [9 August 2004].

Levy, Stephen. *Hackers: Heroes of the Computer Revolution.* Garden City, NY: Anchor Press/Doubleday, 1984.

Lewis, Tom. *Empire of the Air: The Men Who Made Radio.* New York: HarperCollins Publishers, 1991.

Licklider, J. C. R., and Robert Taylor. "The Computer as a Communication Device." *Science and Technology,* April 1968, 40.

Lilienfeld, Robert. *The Rise of Systems Theory.* New York: Wiley, 1978.

Lovelock, James. *Gaia.* New York: Oxford University Press, 1979.

———. *Healing Gaia.* New York: Harmony Books, 1991.

Lyman, Peter, and Hal R.Varian. "How Much Information? 2003." Available [Online] http://www.sims.berkeley.edu/research/projects/how-much-info-2003/ [9 August 2004].

Maclaurin, W. Rupert. *Invention and Innovation in the Radio Industry.* New York: Macmillan, 1949.

Maturana, Humberto, and Francisco Varela. *Autopoiesis and Cognition.* Dordrecht, Holland: De. Reidel, 1980.

McLuhan, Marshall. *The Global Village.* New York: Oxford University Press, reprint edition, 1992.

Milgram, Stanley. "The Small World Problem." *Psychology Today,* May 1967, 60–67.

Mitchell, William. *City of Bits: Space, Place, and the Infobahn.* Cambridge: MIT Press, 1994.

Moravec, Hans. *Robot: Mere Machine to Transcendent Mind.* Oxford: Oxford University Press, 1998.

Morse, Edward Lind. *Samuel F. B. Morse: His Letters and Journals.* Boston: Houghton Mifflin Co., 1914.

Naughton, John. *A Brief History of the Future: The Origins of the Internet.* London: Weidenfeld and Nicolson, 1999.

Navasky, Victor S., and Christopher Cerf. *The Experts Speak: The Definitive Compendium of Authoritative Misinformation.* New York: Villard, 1998.

Negroponte, Nicholas. *Being Digital.* New York: Vintage Books, 1995.

Nelson, Mark R. "We Have the Information You Want, But Getting It Will Cost You: Being Held Hostage By Information Overload." Early 1990s article carried on the ACM site. Available [Online] http://www.acm.org/crossroads/xrds1-1/mnelson.html. [9 August 2004].

Orwell, George. *Nineteen Eighty-Four.* London: Secker & Warburg, 1949.

———. *A Collection of Essays.* New York: Harcourt, 1970.

Oslin, George P. *The Story of Telecommunications.* Macon, GA: Mercer University Press, 1992.

Parsons, Frank. *The Telegraph Monopoly.* Philadelphia: C. F. Taylor, 1899.

Partridge, Craig. *Innovations in Internetworking.* Norwood, MA: ARTECH House, 1988.

Porter, Glenn. *The Rise of Big Business, 1860–1920.* Arlington Heights, IL: Harlan Davidson, 1992.

Poster, Mark. *The Second Media Age.* New York: Blackwell, 1995.

Postman, Neil. "Informing Ourselves to Death." 1990 speech for the German Informatics Society. Available [Online]: http://www.eff.org/Net_culture/Criticisms/informing_ourselves_to_death.paper [9 August 2004].

———. *Technopoly: The Surrender of Culture to Technology.* New York: Vintage, 1993.

Prescott, George B. *Bell's Electric Speaking Telephone: Its Invention, Construction, Application, Modification and History.* New York: D. Appleton, 1884; reprinted by Arno Press, 1972.

Quarterman, John S. *The Matrix: Computer Networks and Conferencing Systems Worldwide.* Bedford, MA: Digital Press, 1989.

Reidenberg, Joel. "Information Flows on the Global Infobahn: Toward New U.S. Policies." In William Drake, ed., *The New Information Infrastructure: Strategies for U.S. Policy.* New York: Century Foundation Press, 1995, 251–68.

Reith, J. C. W. *Broadcast Over Britain.* London: Hodder and Stoughton, 1924.

Rheingold, Howard. *The Virtual Community.* Reading, MA: Addison Wesley, 1993.

Rhodes, Frederick L. *Beginnings of Telephony.* Harper, 1929; reprinted by Arno Press, 1974.

Rosewater, Victor. *History of Cooperative News-Gathering in the United States.* New York: D. Appleton and Company, 1930.

Sale, Kirkpatrick. *Rebels Against the Future: The Luddites and Their War on the Industrial Revolution.* Reading, MA: Longman, 1995.

Salus, Peter. *Casting the Net: From ARPANET to INTERNET and Beyond.* Reading, MA: Addison-Wesley, 1995.

Shiers, George. "Historical Notes on Television Before 1900." *Journal of the Society of Motion Picture and Television Engineers,* March 1977, 129–37.

———. "The Telephone: An Historical Anthology." New York: Arno Press, part of the series *Historical Studies in Telecommunications.* 1977.

Schneider, Stephen, and Penelope Boston, eds. *Scientists on Gaia.* Cambridge: MIT Press, 1991.

Schneier, Bruce. *Secrets and Lies: Digital Security in a Networked World.* New York: Wiley, 2000.

Schubert, Paul. *The Electric Word.* New York: Macmillan, 1928.

Slotten, Hugh R. *Radio and Television Regulation: Broadcast Technology in the United States, 1920–1960.* Baltimore: Johns Hopkins University Press, 2000.

Standage, Tom. *The Victorian Internet.* New York: Walker and Company, 1998.

Sterling, Bruce. *The Hacker Crackdown.* Bantam Books, 1993.

Stoll, Clifford. *The Cuckoo's Egg.* New York: Doubleday, 1989.

———. *Silicon Snake Oil: Second Thoughts on the Information Highway.* New York: Doubleday, 1995.

Taylor, Mark. *The Moment of Complexity: Emerging Network Culture.* Chicago: University of Chicago Press, 2001.

Thompson, Robert Luther. *Wiring a Continent: The History of the Telegraph Industry in the United States, 1832–1866.* Princeton, N.J.: Princeton University Press, 1947.

Thurow, Lester C. *Building Wealth: The New Rules for Individuals, Companies, and Nations in a Knowledge-Based Economy.* New York: HarperBusiness, 1999.

Toffler, Alvin. *Powershift.* New York: Bantam Books, 1990.

Toffler, Alvin and Heidi. *Creating a New Civilization: The Politics of the Third Wave.* Atlanta: Turner Publishing, 1995.

Turkle, Sherry. *Life on the Screen: Identity in the Age of the Internet.* New York: Simon and Schuster, 1995.

Udelson, Joseph H. *The Great Television Race: History of the American Television Industry, 1925–1941.* Tuscaloosa, AL: University of Alabama Press, 1982.

Vinge, Vernor. *True Names.* New York: Dell Binary Star, 1981.

Waldrop, Frank C. *Television: A Struggle for Power.* New York: W. Morrow and Company, 1938.

Waldrop, M. Mitchell. *The Dream Machine: J. C. R. Licklider and the Revolution that Made Computing Personal.* New York: Viking, 2001.

Watts, Duncan J. *Six Degrees: The Science of a Connected Age.* New York: W.W. Norton, 2003.

Weiner, Norbert. *Cybernetics.* Cambridge: MIT Press, 1961.

Williams, Frederick. *The Communications Revolution.* Beverly Hills, CA: Sage Publications, 1982.

Wolf, Gary. *Wired: A Romance.* New York: Random House, 2003.

Wurman, Richard Saul. *Information Anxiety.* New York: Doubleday, 1989.

Yanczer, Peter F. *The Mechanics of Television: The Story of Mechanical Television.* St. Louis: the author, 1987.

Zakon, Robert H. "Hobbes' Internet Timeline v5.0 2000." [Available online]: http://www.isoc.org/zakon/Internet/History/HIT.html [8 July 2004].

Index

About the Author

Janna Quitney Anderson is assistant professor at Elon University. She has directed several major studies for the Pew Internet & American Life Project, building the Internet Predictions Database (www.elon.edu/predictions) and completing an ethnographic study of the use of the Internet by small-town families (www.elon.edu/pew/oneweek/). She is a coauthor of the 2005 Pew Internet report "The Future of the Internet." She has written articles for the *New York Times* News Service, *USA Today*, *Newspaper Research Journal*, *Operant Subjectivity*, and *Advertising Age*. Anderson's 2000 research report on editing and ethics at online newspaper operations received worldwide media attention. She also served as a consultant for the Online News Association's 2001 credibility study. She lives in Elon, N.C., with her husband Daniel, son Tyler, and daughter Kacie.